JAPANESE
HOT SPRING
HOTELS
日本温泉旅馆设计

曲金玲 编

辽宁科学技术出版社
沈阳

图书在版编目 (CIP) 数据

日本温泉旅馆设计 / 曲金玲编 . — 沈阳 : 辽宁科学
技术出版社 , 2017.6
 ISBN 978-7-5591-0182-2

 Ⅰ . ①日… Ⅱ . ①曲… Ⅲ . ①温泉 – 旅馆 – 建筑设
计 – 作品集 – 日本 Ⅳ . ① TU247.4

 中国版本图书馆 CIP 数据核字 (2017) 第 072737 号

出版发行：辽宁科学技术出版社
　　　　　（地址：沈阳市和平区十一纬路 25 号　邮编：110003）
印 刷 者：辽宁新华印务有限公司
经 销 者：各地新华书店
幅面尺寸：230mm × 290mm
印　　张：52
插　　页：4
字　　数：340 千字
出版时间：2017 年 6 月第 1 版
印刷时间：2017 年 6 月第 1 次印刷
责任编辑：于　芳
封面设计：李　莹
版式设计：李　莹
责任校对：周　文

书　　号：ISBN 978-7-5591-0182-2
定　　价：398.00 元

编辑电话：024-23280367
邮购热线：024-23284502
E-mail: 1207014086@qq.com
http://www.lnkj.com.cn

Contents 目录

Shimoda Yamatokan
下田大和馆

Location: 2048 Kisami, Shimoda, Shizuoka Prefecture, 415-0028, Japan

Spa Name: Shimoda Onsen

Area: 17,284 m²

Quantity of Guest Room: 60

Quantity of Spa Facilities: 8

地址：日本静冈县下田市吉佐美2048

温泉名：下田温泉

面积：17284平方米

客房数：60

温泉浴池数：8

Located in Tatado Beach, Shimoda, Shizuka Prefecture, Shimoda Spa perfectly combines Japanese style with natural conditions of the surrounding mountains and sea, making it a cosy and comfortable bathing paradise. Apart from the convenience brought by the modern luxury guest rooms, open-air bathing pools and the dinning rooms, the guests can also indulge themselves in the natural environment with bamboo forest, pine trees, reed mat and Shigaraki Ceramic – the dramatic and organic amalgamation of spa bath and nature, bringing them to the pure nature. With the unique modelling of light and shadow and its transparent visual effect, the design of the spa integrates the beauty of Tatado Beach and the rolling hills, creating a surreal world.

All guest rooms are set within villas with terraces. Guests can overlook the blue sea, whether from the guest rooms or the outdoor bathing pool or even the dinning area.

The fourth floor features the large bathroom named "Paradise View" in which the guest have a bath facing the sea in the ocean-front Western-style large bath and the open-air bathtub. All guests can indulge themselves in all the facilities here upon arrival to departure.

The guests can enjoy the characteristic seafood feast "Umi No Sachi" and "Yama No Sachi" in the guest room or in the dinning room, or even enjoy the charcoal fire barbecue in the dinning room.

下田大和馆温泉位于静冈县下田市的多多户海滨。温泉的设计理念是依据和式风格，通过多多户海滨依山傍海的自然条件巧妙结合而打造出的洗浴"天堂"。游客除享受现代豪华客房、露天浴池及餐厅带给他们的便捷外，还可以尽情享受竹林、松树、芦席和"信乐烧"陶瓷等自然素材构成的原始自然环境，让游人充分体验一种返璞归真的感受，使温泉洗浴与大自然相映成趣，浑然天成。温泉的设计大胆采用了通透的视觉效果和独特的光影、造型，将美丽的多多户海滨和起伏山峦融为一体，创造出一个梦幻般的世界。

客房全部为带有露台的别墅。无论从客房，还是露天浴池及餐厅，游人所到之处均可以远眺碧蓝的大海。

五楼是名为"天堂"的大浴室。里面的日式、西式的大浴池和露天浴盆环抱大海，让游人可以临海而沐。从入住到离开，尽情享用其中的全部设施。

游人可以在客房中、餐厅里享用港町下田特有的丰盛海鲜大餐，如"海之幸"、"山之幸"，也可以在餐厅享用炭火烧烤。

1. The panorama of Shimoda Yamatokan. Located in the natural world of surrounding mountains and sea in southern Izu, the hotel enjoys a panoramic view of the beautiful white Tatado Beach

2. "Kirara", Japanese mountaintop open-air bath

3. The panorama of Shimoda Yamatokan

4. Guest room with private open-air bath which is made of the well-known Japanese Shigaraki Ceramic

1. 下田大和馆外景。矗立于南伊豆，山海环绕的大自然中，美丽的白沙滩"多多户海滨"尽收眼底

2. 和式山顶露天浴池"云母"（Kirara）

3. 下田大和馆外景

4. 有私人专属露天浴池的客房。浴池材料是日本有名陶瓷"信乐烧"

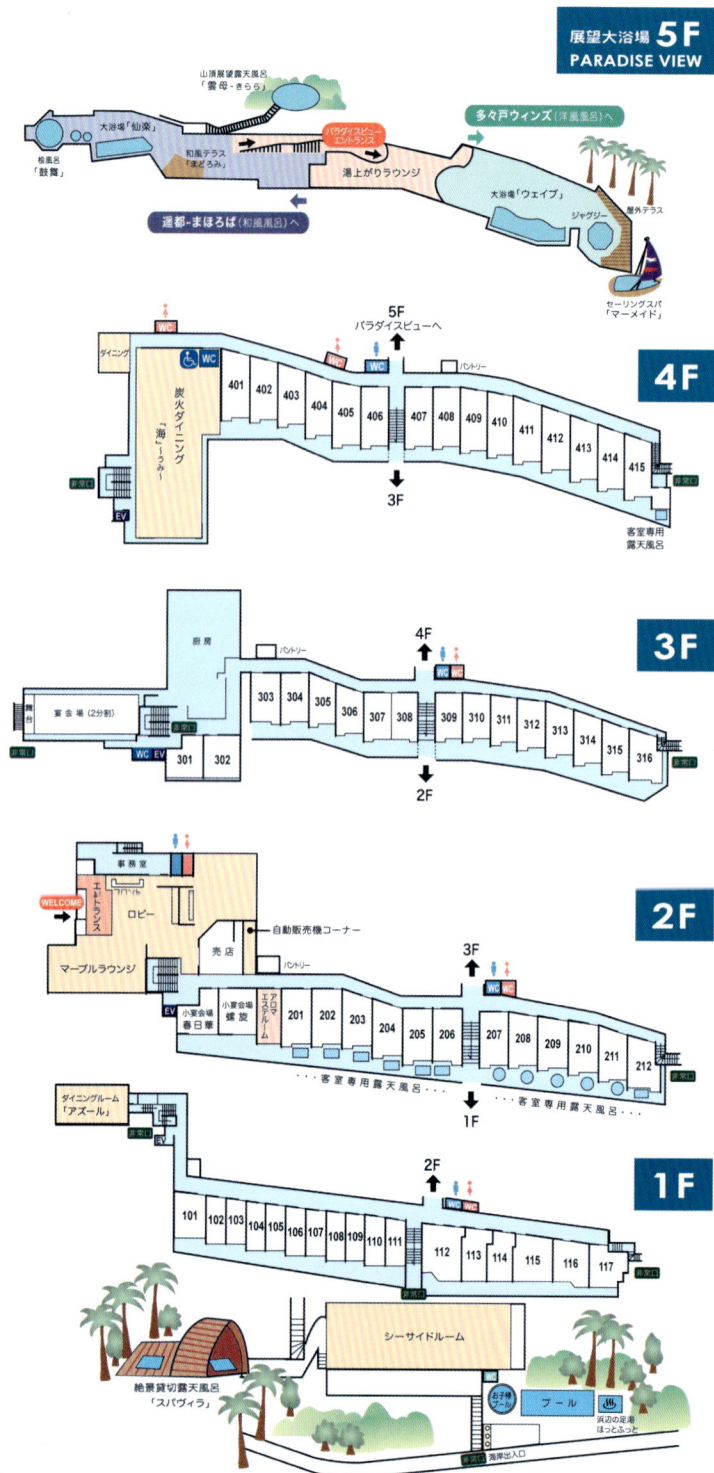

5. plan

5. 平面图

5

3

4

8

6. Private outdoor bath "PINE"

7. Private open-air bath "PINE", featuring the impressive hemisphere-shape roof and the representative pine among the surrounding plants

8. Private outdoor bath "BAMBOO", set within the exotic environment with the surrounding bamboos. The outdoor bath is equipped with special night lighting system.

6. 可出租露天浴池 "PINE"

7. 可出租露天浴池 "PINE"，半球型的房顶和周边绿色植物中具代表性的松树让人印象深刻

8. 可出租露天浴池 "BAMBOO"，营造一种被竹林包围着的感觉，是具有异国风情的优雅环境。露天浴池配备有夜晚特殊照明

9. The most popular guest room 212 and its private outdoor pool equipped with special lighting system

10. Guest room 212

11. Japanese and Oriental style guest room

9. 下田大和馆最受欢迎的212房间及其备有夜晚特殊照明的私人专属露天浴池

10. 212号客房

11. 日式与亚西亚式风格融合的客房

Otsuki Hotel Wafuukan
大月和风馆

Location: 3-19 Higashikaigancho, Atami, Shizuoka Prefecture 413-0012, Japan

Spa Name: Atami Onsen

Area: 2,640 m²

Quantity of Guest Room: 24

Quantity of Spa Facilities: 6

地址：日本静冈县热海市东海岸町3－19

温泉名：热海温泉

面积：2640平方米

客房数：24

温泉浴池数：6

Otsuki Hotel Wafuukan was designed with the motif of "Sanctuary for Adults", therefore it is so appropriate to describe the Japanese spa hotel as a paradise hidden from the city. Located in the downtown area of Atami, Otsuki Hotel Wafuukan Spa boasts the precious tranquility and cosiness. Opposite the vestibule with a lattice door is the spacious room in Japanese style. It also features the Koya Maki Tree open-air bath and the stone bath, made of the 300-year-old superior ancient cypress and the indoor spa bath where guests can catch sight of the green patio through the transparent vertical glass. It was a favourite place of the Tokugawa's. Tourists can go over the interesting history, experiencing wonderful Japanese culture. It is only two minutes' walk to the Atami Sun Beach bathing spot, which makes the hotel more attractive.

The original Japanese teahouse-style spaces, guest rooms in different styles and the Kaiseki Cuisine, which is full of reminiscent mood to Japanese history help guests escape from the hustle and bustle of city life to feel the pure tranquility with time passing by.

Both breakfast and lunch can be enjoyed in the guest room. The Japanese Kaiseki Cuisine select fresh local vegetables and seafood and carefully cooked by the chef.

大月和风馆温泉的设计主题是"成年人的隐匿之处"（隐室）。因此，用"大隐于市"来形容这家纯日式温泉旅馆实在贴切不过了。大月和风馆温泉虽处热海闹市区，却有着难得的宁静闲适。格子门的玄关对面是凝聚着日式风格的宽敞空间。用有300多年树龄的古代上等桧木建造的"高野露天浴池"和"石砌露天浴池"也是大月和风馆的特别之处。透过室内温泉大浴池明亮的立式玻璃，可以看到洋溢着绿色的中庭。日本历史上有名的德川家族也曾在这里享受温泉美景。游人可以在这里寻觅历史，感受浓郁的日本文化。从这里到热海的太阳沙滩浴场只需步行2分钟，这也是大月和风馆备受欢迎的原因之一。

纯日式茶室构造的空间、风格各异的客房以及弥漫着日本怀旧情结的怀石料理，让人们忘却尘世的喧嚣，感受到流淌在身边的每一寸宁静时光。

客人们可以选择在客房内享用早餐、晚餐。这里的日式怀石料理选用当地特产的新鲜蔬菜和海鲜，经过厨师长精心制作而成。

1. The night scene of the vestibule

2. The corridor permeated with Japanese culture

3-4. The Koya Madi Tree spa bathroom – the luxurious independent spa bath made of the ancient cypress which is more than 300 years old

1. 玄关夜景

2. 渗透着日式文化风格的走廊

3-4. 高野露天温泉浴室，使用有300年树龄的古代桧木建造，奢侈的独立式温泉浴室

5

5. The Koya Madi Tree spa bathroom, providing Fujisawa no Yu that exhumed from its own properties, once enjoyed by the Tokugawa family, is one of the famous spas in Japan.

6. The indoor spa bath where the guests can catch sight of the green patio through the transparent vertical glass

5. 高野露天温泉浴室，自家掘出的"藤泽之汤"，日本历史上有名的德川家族曾在此享用温泉美景，是日本有名的温泉之一

6. 室内温泉大浴池，可以透过明亮的立式玻璃，看到洋溢着绿色的中庭

7. The luxurious lounge surrounded by the bamboo garden and pond, outpouring tranquility from its dynamic design

8. The guest room with open-air spa

7. 奢华的大厅休息室被竹园和水池包围，动中显静

8. 带有露天温泉的客房

9

9. Japanese guest room decorated in traditional Japanese colours

10. Guest room in both Japanese and European style, whose transparent design affords a direct view of the garden

9. 日式客房，彰显日式传统色彩搭配风格

10. 集日式和欧式风格于一体的客房，通透的设计，一眼即可看到庭园美景

Seizanyamato

青山大和温泉旅馆

Location: 203 OkaIto-City Shizouka Prefecture, 414-0055, Japan

Spa Name: Ito Onsen

Area: 8,220 m²

Quantity of Guest Room: 42

地址：日本静冈县伊东市冈203

温泉名：伊东温泉

面积：8220平方米

客房数：42

Nestled on the serene and amicable hillside in Ito City, Shizuoka Prefecture, it is an up-to-date Japanese hotel. Taking advantage of the unique topography of Ito, the designer created a paradise on earth, connecting mountains and rivers amidst the overlapping peaks and knolls with use of natural materials such as bamboo, wood, reed, tatami, sliding window and screen. The Japanese architecture technology and wisdom demonstrated here make guests intoxicated in the comfortable and cosy spa.

Seizanyamato provides private open-air baths and rentable spas service. It boasts 42 rooms in total, among which there are 5 high-grade guest rooms with private open-air pool and 2 suites with open-air baths. All the spa facilities, including separate large spa pools for men and women, the outdoor spa bathing spots and the exclusive hourly bathtubs, adopt the water circulating system. The guests can lie in the spa admiring the sparkling stars in the sky at night and overlook the sea far away during daytime with the rippling sound of the running spring nearby .

There are also such supporting facilities as stores, beauty salons, recreation rooms, karaoke clubs and lounges in which you are free to surf the internet. Here serves the custom-made dish menu updated every month, among which the most popular one is

the elaborate Kaiseki Cuisine cooked with exquisite ingredients, featuring Seizanyamato flavour.

There are three choices respectively for drinks, entrée and dessert in the Japanese cuisine called "Seizan Kaiseki Ryori", the ones can be chosen in accordance with personal taste.

青山大和温泉旅馆坐落于静冈县伊东市一处安静的小山丘上。这是一个新潮现代的日式风格旅馆。设计者利用伊东特殊的地势，采用最自然的素材，如竹、木、苇材、榻榻米、拉窗、屏风，在层峦叠嶂的山林中，巧妙构建山水相连的人间仙境，让游人尽享温泉带给他们的舒适和安逸，彰显日本建筑技术和智慧 。

青山大和温泉旅馆客房提供私人专属露天浴池和包租温泉服务，共有42间客房。其中5间为配备了露天浴室的高级客房，2间为带露天浴室的套房。此外还有男宾、女宾专用的温泉大浴场、露天温泉浴场、定时专用的浴盆。所有温泉设施均采用循环流水方式。客人可以躺在温泉里看夜晚满天繁星，亦可在白天听着温泉流动远眺碧蓝的大海。

青山大和馆配有便利店、美容室、娱乐室、卡拉ok室和可免费上网的休息室。这里的食谱每月更新，而最受顾客欢迎的是食材考究、精心制作的青山风味怀石料理。

青山大和的和式料理"青山会席"从酒水、主菜到甜点各有3种，可以根据个人口味选择。

1. The night view of the façade – one corner of the colourful and dazzling yard in the dim light of night

2. The overall view of Seizanyamato

3. The cosy and comfortable lobby

4. The large bath "Tsuki no Yu". At "Tsuki no Yu", you can enjoy a far view of the Ito streets and the water and sky of one hue of Sagami bay.

1. 外观夜景。夜色环抱，鎏光溢彩的庭园一角

2. 绿色环绕的青山大和馆全景

3. 温馨舒适的前厅

4. "月之汤"，可以望到伊东的街道和天水一色的相模湾。汨汨流动的温泉水不停注入到宽大的浴盆里

5. The night view of the vestibule. Blooming Oshima Sakura can be seen around the vestibule in February every year.

5. 玄关夜景，每到2月大岛樱花在玄关四周开放

3

4

6. The open-air bath "Kaze no Yu", covered with mist and water, makes you feel like standing high in the clouds.

7. The open-air bath "Hoshi no Yu" is an open-air stone spa in the face of mountains and woods, running over with the amorous feelings of mountains. The guests can enjoy the sparkling stars all over the sky at night, lying comfortably in the spa.

8. Teahouse, where you can enjoy the cosy space by overlooking the green view through the large window while leisurely tasting the fragrant flavour of tea

6. 露天温泉"风之汤"里，水雾绕缭之时，人如同身处云端

7. 露天浴池"星之汤"是面向山林，洋溢着山野风情的露天石砌温泉。夜晚，慵懒地躺在温泉里，仰头就可以欣赏到满天的星斗

8. 饮茶俱乐部，透过大扇窗一边眺望庭园的绿，一边品尝清茶，慢慢地享受现有的一切

9. The lounge for after-bath, providing various drinks such as pure water and purple Perilla juice which are prepared in the changing rooms which radiate the unique flavour of wood. The guests can also see the green plants in the patio, while regaining physical strength.

10. The beauty salon "Kirakira", where the guests enjoy the massage with natural vanilla essential oil amidst the vanilla smell permeating the inside space

11. The standard guest room, equipped with natural-material-made tatamis, sliding windows and screens, showcasing the Japanese architectural technology and wisdom

9. 沐浴后休息，散发着特有木香的更衣室里，备有伊豆的天然水、紫苏果汁等冷饮。从这里还能看到中庭的绿色植物，让客人们在绿色中慢慢恢复失去的体力

10. 美容室"煌辉"，在香草气息中体验纯天然香草精油的按摩服务

11. 标准客室采用天然材料制成的榻榻米、拉窗、屏风等，彰显出日式建筑技术和智慧，且风格多样

Hotel MICURAS
MICURAS温泉旅馆

Location: 3-19, Higashikaigancho, Atami, Shizuoka Prefecture, 413-0012, Japan

Spa Name: Atami Onsen

Area: 1,320 m²

Quantity of Guest Room: 62

Quantity of Spa Facilities: 4

地址：日本静冈县热海市东海岸町3－19

温泉名：热海温泉

面积：1320平方米

客房数：62

温泉浴池数：4

Hotel MICURAS is located in Sagami Bay, adjacent to Atami beach. Taking advantage of the beautiful beach landscape of Sagami Bay, the designer created sea-view spa pools. The concept was inspired from illimitable sea & sky, and employed blue and white as the main furnishing colours. MICURAS incorporates the bath and wellness services, which had attracted all-aged customers. The spa service, which had been focusing on health care as its main service item, has won good reputation among tourists of all age group, especially the female guests, since its opening in January, 2007. The most brilliant touch goes to the panoramic open spa, where guests can enjoy different sceneries from the ocean front spa which has a panoramic view of the whole peninsular, to the wonderful neighbourhood and night scene of the sand beach, which can be seen directly by lying in the spa bathing pool. In sunny days, an overall view of the whole peninsula can be enjoyed in the spa where bathing seems like swimming in the sea.

MICURAS温泉旅馆位于相模湾，毗邻热海太阳沙滩浴场。借助相模湾海域的美丽风光，设计者打造出开放式海景温泉浴池，构思体现海天相连的恢弘气势，以蓝、白色为装饰主色调。MICURAS集洗浴、疗养于一体，吸引不同层次的消费者。自2007年1月营业以来，原本以疗养为主要服务项目的温泉洗浴受到各年龄段游客的一致好评，尤其深受女性客人青睐。MICURAS设计最得意之作是开放式全景温泉，在这里，游客可以观赏到不同的景致：既有将相模湾的初岛、大岛景色一览无余的敞开式海景温泉，也有躺在温泉浴池里就可以看到的热海美丽街区和沙滩夜景。天气晴朗时整个半岛尽收眼底，身在露天温泉却如同畅游于大海中一般。

1. Hotel MICURAS on the Atami Beach in warm sea wind

2. Open ocean front communal spa bath III, where one can catch sight of the beautiful neighbourhood and the night scene of the Atami sun beach

3. Open ocean front communal spa bath II. It has a transparent view from the interior to the outside. The hot spring here can improve your skin to be smoother and gentler.

4. The corridor of men's exclusive lounge

1. MICURAS旅馆矗立于热海海滨沙滩边上，图为温暖的海风吹拂下的MICURAS旅馆

2. 敞开式海景大浴池3，躺在浴池里就可以看到美丽的街区和沙滩夜景

3. 敞开式海景大浴池2，从室内部分望向室外。这里的温泉水质可使肌肤更加光滑柔顺

4. 男士专用休息室走廊

5. Open-air ocean front large spa pool with a commanding view of Sagami Bay, Hatsushima and Oshima

5. 敞开式海景温泉大浴池，相模湾的景色一览无余，初岛、大岛也尽收眼底

6. The luxurious suites for honeymoon, open-view pool provides a panoramic view of the seascape

7. The panoramic view of the restaurant, where the hazy candle light combined with the leisurely music will accompany you for dinner time

6. 豪华蜜月套房，有尽享海景的温泉浴池

7. 餐厅全景，朦胧的烛光和悠扬的音乐将与你共进晚餐

8. The overall scene of the MICURAS Café, which locates along the seaside, offering a cosy environment for enjoyable afternoon tea

9. The guest room with indoor bath pool

8. 咖啡厅全景。位于沿海大道的MICURAS咖啡厅，让你尽享惬意的下午茶时光

9. 带室内温泉浴池的客房

10

10. The MICURAS Suite with spacious living room

11. The luxurious ocean-view guest room. There are luxurious private ocean front bath pools, high-tech beauty and spa facilities. The spacious living room, bedrooms as well as the beautiful seascape make you fully intoxicated in the incomparable holiday

12. The night scene of the MICURAS Suite, with ocean-view spa pool

10. MICURAS套房宽敞的起居室

11. 豪华海景客房。客房内设有豪华的私人海景浴池以及高科技美容疗养设施。宽敞的客厅、卧室、美丽的海景风光让你尽享无与伦比的假日

12. 套房夜景。MICURAS套房的卧室备有海景温泉浴池

The Prince Hakone Resort
箱根皇家王子度假村

Location: 144 Motonhakone, Hakone-machi, Ashigara-shimo-gun, Kanagawa, 250-0592, Japan

Spa Name: Lake Ashi Takokawa Onsen

Area: 94,500 m²

Quantity of Guest Room: 138 (Main building 94; Annexe 44)

Quantity of Spa Facilities: 4

地址：日本神奈川县足柄下郡箱根町元箱根144

温泉名：芦之湖蛸川温泉

面积：94500平方米

客房数：138（本馆94、别馆44）

温泉浴池数：4

The Prince Hakone Resort has two famous spas in Hakone: Kohan Spa and Yumedono Onsen with hot spring coming from Takokawa Onsen, one of the top seventeen spas in Hakone. The Lakeside Spa in the Prince Hakone boasts the natural advantage of spas and the adjoining mountains and rivers, enjoying a commanding view of Mount Fuji and adjacent Lake Ashi. Mount Fuji endows the spa with enchanting glamour in all seasons of a year. The Prince Hakone Resort is the flagship store of the Prince Grand Hotel and Resorts Group. Guests here can enjoy the comfortable holiday time and the infinite space in harmony with nature, intoxicating in such warm embrace of Lake Ashi and Hakone mountains and forget to leave. The open-air spa against the natural beauty of lakes and mountains provides the guests with the delight in the gentle sunshine and the lake landscape. Guests can totally soothe their mind and body by overlooking Mount Fuji from Lake Ashi. The spa here is rich in the natural elements such as calcium, sodium, sulfate and chloride, which have the significant effects in curing neuralgia, myalgia and arthralgia, and moreover can relax the muscles, stimulate the blood circulation and alleviate travelling fatigue.

箱根皇家王子度假村在箱根有两个著名的温泉——湖畔温泉和梦殿温泉，它们的源泉都来自蛸川温泉，名列著名的“箱根十七泉”。箱根皇家王子大饭店的湖畔蛸川温泉，以温泉与山水相连的自然优势，坐享富士山风光，依傍芦之湖秀色。富士山为温泉的春夏秋冬披上了迷人的光彩。箱根皇家王子度假村是王子大饭店和度假村集团的旗舰店。在这里，游客可以享受与大自然和谐相融的写意时刻和无限空间，环抱于芦之湖和箱根群山之中，陶醉于此而流连忘返。游客们可以在映衬着湖光山色的露天温泉湖畔中，尽享柔和的阳光湖景。从芦之湖畔远望富士山，身心将无拘无束地舒展开。箱根皇家王子度假村温泉泉质富含钙、钠、硫酸盐和盐化物等，有治疗神经痛、肌肉痛和关节痛等功效，又可以舒筋活血，缓解旅途疲劳。

1. Fuji Mount at the Hakone lakeside

2. Fuji Mount behind Nakaniwa Lake with implicit and charming beauty just like a shy and pretty girl playing pipa

3. The open-air spa adjacent to lakes and mountains

4. The indoor spa pool

1. 箱根芦之湖的富士山

2. 中庭湖后的富士山犹抱琵琶半遮面

3. 依傍湖光山色的露天温泉

4. 室内温泉浴池

5. The exterior view of the Prince Hakone Resort

5. 箱根皇家王子度假村外观景色

3

4

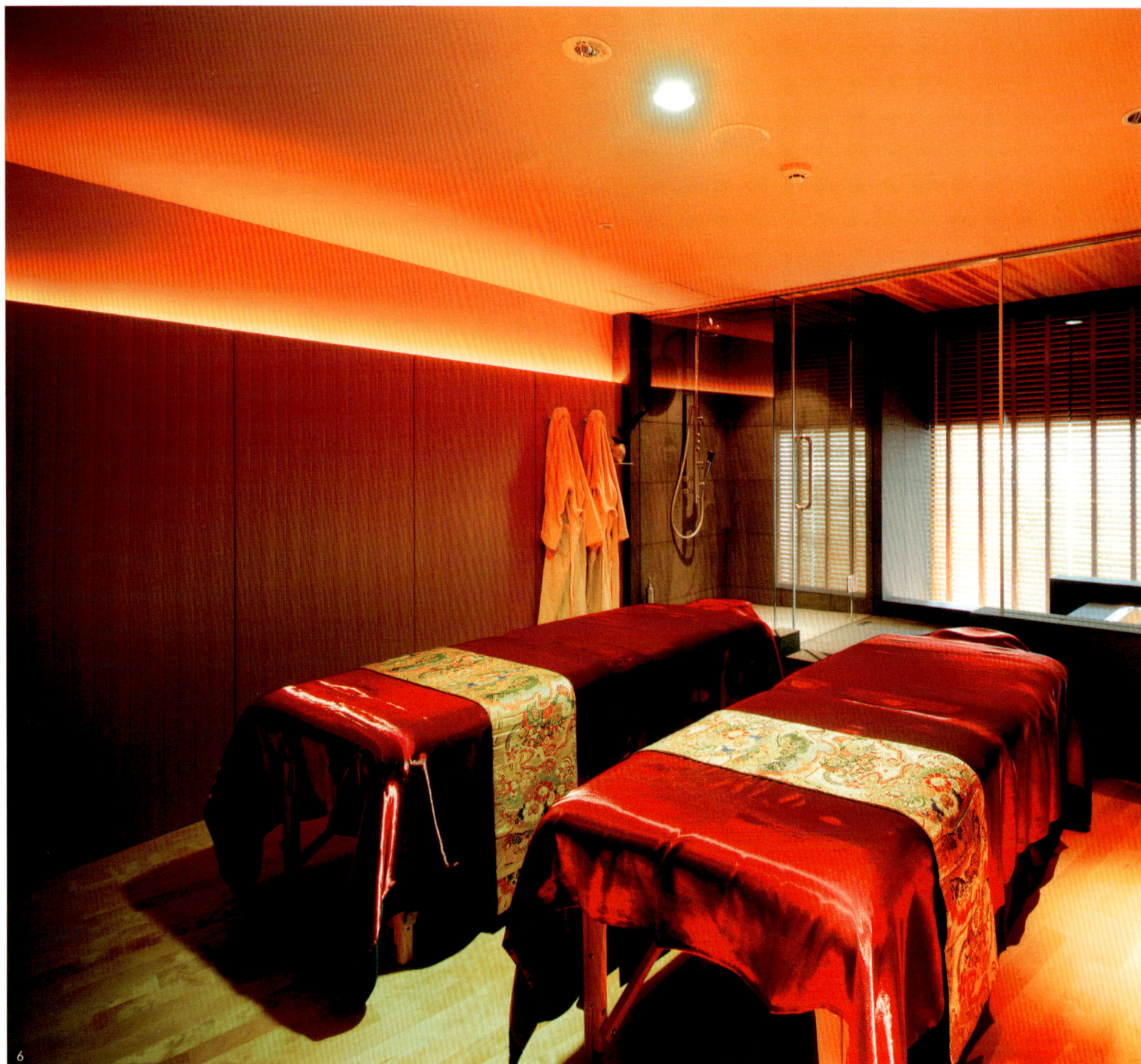

6. The massage room

7. One of the guest rooms, boasting gentle sunshine and lake view

8. The guest room in the annexe

6. 按摩室

7. 客房之一，尽享柔和的阳光湖景

8. 别馆客房

Ootaki Hotel

大泷饭店

Location: 750 – 1 Miyakami, Yugawara-machi, Ashigarashimo-gun, Kanagawa Prefecture, 259-0314, Japan

Spa Name: Yugawara Onsen

Area: 20,955 m²

Quantity of Guest Room: 43

Quantity of Spa Facilities: 5

地址：日本神奈川县足柄下郡汤河原町宫上750-1

温泉名：汤河原温泉

面积：20955平方米

客房数：43

温泉浴池数：5

Ootaki Hotel possesses one of the largest volumes of hot spring water in the Yugawara hot spring resort and is located adjacent to the picturesque Fudou-taki(Fudou Waterfall). As one of the rare natural onsen resorts with rich soil and resource in suburban Tokyo, it boasts comprehensive range of hot spring facilities including open-air bath with views of surrounding landscape, hot spring swimming pool that can be used all year round, swimming-wear-required rooftop bath and foot onsen, while it takes most pride in the open-air baths which not only command a view of the famous Fudou Waterfall but enjoy the most plentiful hot spring volume. It features seasonal Japanese Kaiseki Ryori created with local vegetables and sea food. Hotel has installed wheelchair slope at the entrance and lift capable for accommodating wheelchairs. Hotel has also established barrier free rooms and handrails in open-air bath.

大泷饭店位于日本著名温泉之乡"汤河原温泉"，与汤河原温泉名胜景点"不动瀑布"相邻，是东京近郊少见的自然风情浓郁之地，所在地泉水自然喷涌而出。露天观景浴池是大泷酒店引以为豪的，不仅能够眺望到著名的"不动瀑布"，而且拥有汤河原温泉最大的出水量，丰沛充盈。温泉设施齐备，拥有全年开放的温泉游泳池、屋顶野外露天大浴池（可穿着泳衣前往）、泡脚温泉等，游客可以随时随地享受温泉带来的舒适和惬意。酒店提供季节性浓郁的会席料理，精选当地的山珍海味，精心烹制而成，味道纯正、好评如潮，是酒店的又一大特色。屋顶庭园里安置着大正时代建造的广部观音，环境优雅，最适合在此闲庭信步。酒店采用无障碍设计，门前的斜坡、轮椅平台电梯以及露天温泉里的扶手，尽显人性化关怀。无障碍客房也已全面建成。

1. The façade of the eight-storey building with hot spring water supply for all the rooms

2. The open-air spa bath "Kanontaki no Yu", swimwear-required for both men and women

3. The foot bath "Kanon"

4. The open-air spa bath "Kanontaki no Yu"

1. 8层建筑外观，全部房间都有温泉水供应

2. 露天温泉浴池"观音泷之汤"，男女共用

3. 足浴场所，名为"观音"

4. 露天温泉"观音泷之汤"

5. The surrounding attraction of the famous "Fudou-taki" of Yugawara Onsen

5. 汤河原温泉名胜景点"不动瀑布"及周边环境

3

4

6. The outdoor spa of "Taki no Yu" 6. "泷之汤"的露天温泉

7. The outdoor spa of "Hotaru no Yu" 7. "萤之汤"的露天温泉

8. The indoor pool "Taki no Yu" 8. "泷之汤"室内大浴场

9. The indoor pool "Hotaru no Yu" with the top-class hot spring from Yugawara Onsen, whose rich volume brims the Kitakami stone bath

10. The exlusive open-air spa in the guest room

11. The hot spring swimming pool, which is available all year round

9. 室内大浴池"萤之汤"，汤河原温泉最优质的温泉，汩汩的源泉始终注满木曾石头浴槽

10. 客房专用露天温泉

11. 温泉游泳池，一年四季均可使用

12

12. One of the guest rooms with open-air bath

13. The standard Japanese-style guest room

14. One of the Western-style guest rooms

12. 带专用露天温泉客房

13. 标准和式客房

14. 西式客房

13

14

Hakone Washintei Hougetsu
和心亭丰月温泉旅馆

Location: 90-42 Hakone, Hakone, Asigarasimogun, Kanagawa Prefecture, 250-0522, Japan

Spa Name: Lake Ashi Onsen

Area: approx. 4,000 m²

Quantity of Guest Room: 15

Quantity of Spa Facilities: 4

地址：日本神奈川县足柄下郡箱根町元箱根90-42

温泉名：芦之湖温泉

面积：约4,000平方米

客房数：15

温泉浴池数：4

The name of the hotel Hakone Hougetu comes from the beautiful full moon views in Hakone Mountain with the central theme of moon. The ideal location on the mound enables guests to admire the vivid song of birds and overlook the stunning seasonal landscape of Lake Ashi. The interior decoration everywhere set off the poetic imagery of full moon such as the "Fugetu no Yu", "Tsukimi no Yu" and "Deai Tsuki no Yu", pursuing a kind of simple and elegant mood and spirit, which implies the dream for the natural elements in Japanese culture whilst presents the guests the comfortable and cosy private space. Whether in the open-air common bath or the private exclusive bath, guests can enjoy themselves in high-quality sulfurous hot spring of Lake Ashi Onsen while admiring the splendid starry sky.

Equipped with "Hori–Kotatsu" (low heated table with legroom built into the floor), the spacious rooms with varied layouts create a sense of openness. The Japanese Ryotei restaurant with private rooms has been serving seasonal cuisine with monthly updated menu since its opening. Guests can enjoy the pleasure of tasting the exclusive cuisine cooked with local ingredients and admiring the stunning scenery of Hakone at the same time.

和心亭丰月温泉旅馆依据箱根山"满月"时的魅力而取名"丰月"，设计理念也是以"月"为主题。温泉旅馆建在高岗上，游人在温泉沐浴的同时，可以耳闻鸟啼声声，俯瞰芦之湖四季美丽景致。馆内装饰设计上处处衬托"满月"般的意境，如"风月之汤"、"见月之汤"、"逢月之汤"，从而追求一种淡雅而温和的心情，体现出日本文化对自然的追求和向往，给游客提供一种舒适惬意的私密空间。不论是露天大浴池还是私密的包租温泉浴池，都能够让游客尽情享受芦之湖温泉的优质硫黄温泉。

客房都配有炕式暖桌，室内宽敞、格局各异，有开放感。从客房可一览芦之湖美景。带包间的亭式日本料理餐厅，自开业以来，一直提供充满季节感的美食，并且每月更新菜肴。游客们既可以品尝到当地食材独创的料理，亦可同时观赏箱根的美景。

1. The hotel with the central theme of "Tsuki", attracted by the angelic image "Hougetu" in Hakone

2. The typical Japanese hotel, named in odd number

3. The "Yukemuri Tsuki" indoor bathing pool with the natural screen through large windows

1．被箱根山的"丰月"（满月）魅力吸引，以"月"为主题的旅馆

2．以奇数而造的纯和式旅馆

3．"见月之汤"室内浴池的大落地窗如同自然的大屏幕一般

4. The "Tsukimi no Yu" outdoor bathing pool

4．"见月之汤"室外露天浴池

5. The "Fugetu no Yu"

6. The "Deai Tsuki no Yu" bathing pool in the form of a bathtub made of high-grade cypress

5. "风月之汤"

6. "逢月之汤"高级桧木制成的浴盆型浴池

7

7. The modern-style chartered bath, focusing on the private space

8. The indoor bath pool presents a different feeling by bathing in the rounded bathing pool and admiring the moon and stars through the window

9. The design element demonstrates the owner's adoration to moon

7. 注重私人空间，现代风格的包租浴池

8. 浸泡在室内浴池的圆形浴池，透过大玻璃窗赏月观星，别有一番情趣

9. 设计上也可见馆主对月的喜爱

10

11

12

10. The corridor to the reception hall whose design focuses on the creation of the Japanese beauty with natural ornaments

11. The reading room for guests' leisure time, equipped with massage armchairs

12. The reception hall where you can catch sight of the stunning scenery of Lake Ashi and the surrounding landscapes

10. 通接大厅的走廊，设计上注重利用自然饰物打造出"和"的空间美

11. 客人专用书房，藏有多种与箱根有关书籍，并配有按摩椅

12. 接待大厅，在此小憩，可观赏箱根芦之湖和周围美丽景致

13. The Japanese-style guest rooms which command a panoramic view of Lake Ashi

14. Various room layouts according to the quantity of guests

15 The private room in the restaurant

13. 从日式客房可以一览芦之湖美景

14. 房间格局因游客数而布局多样

15. 餐厅包间

Gora Tensui
强罗天翠温泉馆

Location: 1320-276 Gora, Hakone, Ashigarashimogun, Kanagawa Prefecture, Japan

Spa Name: Hakone Gora Onsen

Area: 2,775 m²

Quantity of Guest Room: 14

Quantity of Spa Facilities: 4

Other Facilities: "Ashiyu" Café; Rock Base Bath; Sauna and Esthetics Salon

地址：日本神奈川县足柄下郡箱根町强罗

温泉名：箱根强罗温泉

面积：2775平方米

客房数：14

温泉浴池数：4

其他设施：足浴咖啡酒吧、岩盘浴、桑拿、美容室

It takes only one minute to walk to Gora Tensui after getting off from the mountain train at Gora Station. Incorporating the modern and traditional Japanese architectural style to the whole design, the design of the guest room emphasises the privacy as well as the ornamental characters. The ideal location affords six sightseeing choices. The guests can catch sight of the grand occasion of the Hakone traditional culture ceremony, "Daimonji Matsuri" and overlook "Mt. Myojogatake" in Hakone Izu National Park. It also boasts the elegant surrounding landscape such as the splendid flower sea of plum blossom and sakura in early spring, the fragrant blossoming lilac in midsummer, the fiery maple leaf layers all over the mountain in early autumn and the snow-coated tress in winter.

Acidity defines the quality of hot springs here, and there are milky white salt elements with acidity in the water of "Bansui no Yu" and "Sansui no Yu". This kind of element, blended with the alkalinity of human cells, can develop the effect of relieving fatigue and defusing the vivotoxin. The exclusive bath and the rock base bath are free. There are also irregularly scheduled culture exhibitions of each and every kind.

从箱根登山铁路"强罗站"下车步行1分钟即可到达强罗天翠温泉馆，交通便捷。酒店融合了现代和传统日式建筑风格。客房设计除注重私密感外，更注重了观赏性。因地理位置优越，共有6种观光类型可供游客选择。全部客房均可以观赏到箱根传统文化祭——"大文字祭"的盛况，并且可以远眺到富士箱根伊豆国立公园的"明星之岳"。周边风景秀美宜人：初春，梅花、樱花浩如花海；仲夏，紫丁香花香气袭人；入秋，火红枫叶层林尽染；隆冬，满树枝头银装素裹。

这里的温泉水质呈酸性，其中"万翠之泉"和"山翠之泉"色呈乳白，富含酸性及盐化元素，与人体细胞所含碱性相交融，具有解除疲劳，化解体内毒素的特殊功效。馆内提供免费浴池和岩盘浴，并且不定期举办各种文化展览活动。

1. The night scene of the entrance

2. The contracted open-air bath named "Myojogatake no Yu" which can be used without reservation. Somma in Hakone is right under your nose when you bathe in the open-air bath.

3. The night scene of the natural hot spring "Bansui no Yu"

4. The open-air spa of the Japanese guest room "Suisei"

1. 入口夜景

2. 租赁私人专属露天温泉"明星之泉"无须事先预约。浸泡在温泉里,箱根外轮山近在眼前

3. 露天温泉"万翠"夜景

4. 日式客房"翠星"的露天浴池

5. The vestibule with a direct access to the most representative "Ashiyu" Café

5. 玄关直接通往馆内最具代表性的足浴咖啡酒吧厅

6-7. The "Ashiyu" Café "Mastuyoi". You can enjoy the first welcome drink as complimentary while having your weary feet sank in the hot spring.

6-7. 足浴（温泉足浴）咖啡酒吧厅"待宵"，可坐在这里浸泡疲倦的双脚，并享受赠送饮品

8. The Japanese and Western-style guest room "Suien" with private open-air bath, creating the modern and traditional atmosphere with the amenities of the high-grade "Ryukyu Tatami" and the low bed design

9. The standard guest room with the high-grade "Sulphur Tatami", totalling 12 tatami mat (20 square metres), together with a small-size living room and a shower

10. The Japanese guest room named "Suikei" totalling 12 tatami mat (20 square metres) plus 7.5 tatami mat (12 square metres)

8. 客房"翠苑",带有专属温泉浴池,日式与西式风格相结合,采用高级"硫球榻榻米"和低位床的设计营造现代与传统结合的氛围

9. 日式标准客房,室内采用高级"硫球榻榻米",12帖榻榻米(约20平方米)加小厅和淋浴室

10. 带有专属露天浴池的日式客房"翠景",12帖榻榻米(约20平方米)加7.5帖榻榻米(约12平方米)

Hotel Kajikaso

河鹿庄温泉酒店

Location: 688 Yumoto, Hakone-machi, Ashigarashimo-gun, Kanagawa, Japan

Spa Name: Hakone Yumoto Onsen

Area: about 9,900 m²

Quantity of Guest Room: 74

Capacity: 430 persons

Spa Facilities: Indoor sap pool 2, outdoor spa 8 in 5 types and foot spa

地址：日本神奈川县足柄下郡箱根町汤本688

温泉名：箱根汤本温泉

面积：约9900平方米

客房数：74

定员：430名

温泉设施：大型室内温泉浴池2个、露天温泉5种8个、泡足温泉

It takes only five minutes to walk to Hotel Kajikaso after getting off the train at Hakone Yumoto Station. The designer of the hotel took full advantages of all the spaces of the spa. There is an observatory large communal spa bath named "Seiryuu no Yu" on the roof terrace on the fifth floor and an open-air spa "Soun no Yu" on the first floor. The spa services are available to men and women at different time intervals. There are nine spas in total with rich water volume. The elegant surrounding environment commanding a panoramic view of the stunning seasonal landscape of Hakone Mount, makes guests feel at ease, which is also the main idea of the hotel. On the wooden deck on the first floor, guests can take a rest by having a foot spa and enjoy the natural oxygen bar, while listening to the rippling sound of the running brook. The hotel is also equipped with various public recreational amenities such as Karaoke Hall, Mahjong Hall and Banquet Hall in different sizes. With the good reputation of "Home of Kajikaso's Food", the hotel offers exceptional Japanese cuisine cooked with fresh seafood from Sagami Bay.

There are a large number of tourist attractions around, namely, river fishing, tennis court, golf, hiking, mountain-climbing, mountain vegetables picking, museum, aquarium, botanical garden and art gallery.

河鹿庄温泉酒店从登山铁路"箱根汤本站"下车步行5分钟即可到达。设计者的初衷是将温泉的空间感发挥到极致。六楼屋顶平台备有大温泉浴池"清流之汤"，与二楼的"早云之汤"皆为露天温泉，男女浴池按不同时段交换使用。馆内共有源泉9处，涌泉量丰富。四周环境优雅，可将箱根山四季变幻的美景尽收眼底。让游人感到舒适、放松是这家酒店的服务宗旨。游客在二楼的木凉台可以一边倾听早川的涓涓溪流，一边小憩泡足浴，享受天然氧吧。酒店还配备各种娱乐性公共设施如卡拉OK、麻将室、游戏厅、大小宴会厅等。餐厅素有"河鹿庄美食之乡"的美誉，提供相模湾当地的特色海鲜美食。

周边观光景点及休闲活动：溪钓 、网球场 、高尔夫 、徒步旅行 、登山 、博物馆 、水族馆 、植物园 、美术馆 、采山菜。

1. The façade of the typical Japanese hotel facing Hayakawa

2. The large open-air communal bath named "Soun no Yu". "Soun no Yu" on the first floor is for men exclusively.

3. Outdoor spa pool named "Tsukimi no Yu"

4. The foot bath. The fatigue of the feet is relieved.

1. 面向早川，格调高雅的纯日式旅馆外观

2. 大浴池"早云"，二楼备有男性专用"早云之汤"露天温泉

3. 露天温泉"望月之汤"

4. 足浴

5. The four observatory communal spas "Hoshimi no Yu"

5. "望星四汤" —— 4个展望式露天温泉

3

4

6

7

6. The lounge named "Keiryuu"

7. The lounge of the ante chamber named "Fuga"

6. 休息室 "溪流"

7. 前厅休息室 "风雅"

8. The Japanese guest room with the layout of 20-square-metre interior and an outdoor spa, which has a room capacity of 6 persons, suitable for the family tour

9. The Japanese guest room with a layout of 20-square-metre interior and an outdoor spa, which has a room capacity of 6 persons, suitable for the family tour

10. The Japanese and Western style guest room "Syusuian", 16-square-metre interior, a living room and an outdoor spa facing the stream

8. 日式客房，20平方米室内空间+露天温泉，定员6人，适于家庭旅游时入住

9. 日式客房，20平方米室内空间+露天温泉，定员6人，适于家庭旅游时入住

10. 日式和西式融为一体的客房——"集粹庵"，16平方米室内空间+客厅+面朝溪流的露天温泉

Yushintei
游心亭旅馆

Location: 193 Yumoto, Ashigarashimo-gun, Kanagawa Prefecture, Japan

Spa Name: Hakone Yumoto Onsen

Area: 953.32 m²

Quantity of Guest Room: 10

Quantity of Spa Facilities: 5

地址：日本神奈川县足柄下郡箱根町汤本茶屋193

温泉名：箱根汤本温泉

面积：953.32平方米

客房数：10

温泉浴池数：5

Yushintei is based in the tourist attraction of Yumoto, Hakone, Kanto, Japan. It is only one kilometre away from Hakone Yumoto Staion. Fifteen to twenty minutes' walk along the Skumogawa River towards Tasidoori Onsen Town can lead you to the destination in the deep lane of the hotel street, which endows the spa with the lingering charm, profound and lasting, small but exquisite. It is a two-storey building in wooden structure. The Kiasian on the ground floor with open-air spas are named as "Kisaian Ai", "Kisaian Kohaku", "Kisaian Toki" and "Kiasian Sawarabi". The modern-style design with private open-air spa offers guests a feeling of absolute privacy.

The natural alkaline water of Yushintei Spa can not only relax the muscles and stimulate the blood circulation but exert unique function in maintaining beauty image for the female. The tranquility and the considerate and thoughtful service make guests intoxicated in the ultimate relaxation and enjoyment of splendid holiday time.

游心亭旅馆位于日本关东地区旅游胜地箱根汤本，距离箱根汤本电车站大约1公里路程，沿着须云河，徒步往"泷通"（路名：Takidoori）温泉乡的旅馆街深处走15～20分钟即可到达。由此，温泉也就有了深远悠长、小巧玲珑的韵味。箱根汤本游心亭旅馆为双层木结构建筑物。一楼"季彩庵"设有露天温泉浴池，分别以"蓝"、"琥珀"、"朱鹭"、"早蕨"命名。以现代和式风格为主题设计，配备私人专属露天浴池。

箱根汤本游心亭温泉水为碱性纯天然水，不仅有舒筋活血之功能，对女性美容养颜也有特殊功效。箱根大自然的空幽之感和旅馆细致入微的日式服务，可以让游客完全放松身心，享受假日美好时光。

1. The entrance façade of Yushintei, gentle and cosy

2. The exclusive stone spa, rentable by hour

3. One corner of the vestibule inside, in the simple and gentle style

4. Ryokuin, where the guest can enjoy the green view of the courtyard while enjoying the spa

1. 游心亭入口外观，柔和温馨

2. 可以定时租用、私人独享的石砌温泉浴池

3. 馆内玄关一角，展现简单柔和的设计风格

4. 绿荫之泉，可以一边浸泡温泉，一边享受绿色的庭园

5. The vestibule and the front desk, mostly decorated with Japanese paper, creating an atmosphere of a warm home in the gentle light

5. 玄关和总服务台，多处利用和式纸材进行装饰，再配以柔和的光线，营造出宾至如归之感

3

4

6. The open-air spa bath in the guest room "Kisaian Kohaku"

7. The open-air spa bath in the guest room "Kisaian Ai", a wooden spa pool in subtropical style

6. 客房"琥珀"内的露天温泉浴池

7. 客房"蓝"，配有纯木制亚热带风情的温泉浴池

8. The guest room with Japanese feelings, enjoying the beauty of nature

9. The restaurant and typical wooden desktop in semi-closed style

10. The guest room "Kisaian Kohaku"

8. 和式风情客房，尽享窗外大自然

9. 名为"常磐"的餐厅，纯木工特制半包围式餐桌

10. 客房"琥珀"

9

10

Kamogawakan

鸭川温泉旅馆

Location: 1179 Nishi-cho, Kamogawa City, Chiba Prefecture, Japan

Spa Name: Kamogawa Onsen

Area: 18,915.7 m²

Quantity of Guest Room: 65

Quantity of Spa Facilities: 20

地址：日本千叶县鸭川市西町1179

温泉名：鸭川温泉

面积：18 915.7平方米

客房数：65

温泉浴池数：20

Located in the south of Boso Peninsula, Kanto, Kamogawakan enjoys a warm and pleasant seasonal climate. Facing the Pacific Ocean, Kamogawn is constructed in a simple but elegant Sukiya-zukuri style building structure amidst the surrounding pine trees (Sukiya-zukuri style building structure is a Japanese architectural style, featuring a main building whose façade looks like a teahouse and a courtyard). There are 17 spa pools inside the capacious space including a rooftop footbath "Haruka" commanding a splendid view of the Pacific Ocean. Various types of guest rooms are available here such as the standard Japanese-style, the mixed style of Japanese and Western-style, the ones for lovers and the ones with exclusive open-air baths, whilst the most outstanding guest room goes to the one on the top floor which has the picturesque scenery.

Boso Peninsula has long been crowned as the treasure-house of marine products. Thus, the seafood here is always fresh and delicious. Kamogawakan specialises in serving local seafood which is made of ingredients from Kamogawa, in order to maximise the freshness of the dishes. The open kitchen presents guests a freindly and dynamic feeling.

鸭川温泉旅馆地处日本关东地区房总半岛南侧，这里一年四季气候温暖宜人。旅馆毗邻辽阔美丽的太平洋，温泉周围松林环绕，是简洁美观的数寄屋造式建筑（数寄屋造是一种日式建筑风格，特点是建筑物配有庭园、外观感觉像茶室），共设有17个浴池。旅馆屋顶设有视野开阔的足浴场所，设有各式客房：标准和式风格、日式西式融合风格、情侣式客房等，值得一提的高楼层的客房，拥有如画一般的风景。

房总半岛被誉为日本的海产品宝库，这里的海鲜食材新鲜美味。鸭川温泉旅馆所使用的海鲜食材直接来源于当地的鸭川港，最大限度的保持着美味鲜度。开放式厨房更是给客人一种亲近感和鲜活感。

1. The façade of the onsen hotel in the graceful architectural style

2. The bubble bath for men, an octagon bathing pool with the smell of cypress

3. Enjoying the splendid seascape of the Pacific Ocean while having a foot spa

4. The large bathing pool in men's spa

5. The large bathing pool of women's spa, among which there are three bathing pools with different temperatures that can be chosen according to personal body condition

1. 建筑风格优雅的温泉旅馆外观

2. 男士浴室中的露天气泡浴池，带有柏木温暖气息的八角形浴池

3. 在屋顶一边泡着足汤，一边欣赏太平洋那雄伟壮丽的景观

4. 男士浴室中的大浴池

5. 女士浴室中的大浴池，3个浴池有不同的温度，可以根据身体适应能力选择不同浴池

6. The open-air bath in the guest room, facing the sea

6. 带有专用露天浴池的临海客房

7

7. Massage and therapy hall

8. Standard Japanese suite including one Japanese guest room with 10 tatamis mat (16.2 square metres) and two Japanese guest rooms with 4.5 tatamis mat (7.29 square metres)

9. Standard guest room, mixed Japanese and Western style

7. 按摩疗养室之一

8. 标准日式客房，由10张榻榻米（约16.2平方米）+4.5张榻榻米（7.29平方米）的两个和式房间构成

9. 日式西式结合的标准客房

Amagisou
天城庄

Location: 359 Nashimoto, Kawazu-cho, Kamo-gun, Shizuoka Prefecture, Japan

Spa Name: Oodaru Onsen

Area: 500,000 m²

Quantity of Guest Room: 44

Quantity of Spa Facilities: 37 in 28 types

地址：静冈县贺茂郡河津町梨本359

温泉名：大瀑温泉

占地面积：约50万平方米

客房数：44

温泉浴池数：37（28类）

Sited in the central area of Izu Peninsula in Kanto, Amagisou is crowned as one of the most famous hot springs in Japan. Occupying 500,000-square-metre site, it was built along the Kawazu River, featuring the open-air bath "Kawara no Yu" which commands a panoramic view of the best-known "Otaki", one of the top seven great waterfalls in Kawazu. The stunning scenery of nature, together with numerous grand indoor spas of various kinds throughout the garden, constitutes an unmatched spa kingdom. Besides the picturesque "Kawara no Yu", the ryokan also boasts the 30-metre-deep "Hidden Cave Spa Bath", the metal bathtub heated from beneath with typical Japanese characteristics, "Kodakara no Yu" which may bless you to have many sons and grandsons after bathing in it and "25-metre Hot Spring Swimming Pool" which is available in all seasons. The experiences of viewing flower blossoms in Spring, snow sightseeing in Winter, admiring stars in the sky at night and listening to sounds of the natural fields, being a part of the natural mountains and rivers, make guests enjoy themselves so much as to forget to leave.

There are as many as 28 spa pools such as indoor pools, open-air baths, cave baths, waterfall bath, spa baths (available in winter), stone baths and family baths, which may take guests several days to experience them all.

大瀑温泉天城庄被誉为日本百大名泉之一，地处日本关东区伊豆半岛正中部。沿河津河而建，占地约50万平方米，露天浴池"河原之泉"更是独享河津河七瀑布中最有名的"大瀑"的绝美景观。动静融合的自然美景，再加上庭园内数目众多、形式各异的露天温泉浴池，构成了伊豆独一无二的温泉王国。除了美景尽享的"河原之泉"，还有深度为30米的"秘泉穴浴池"。此外，还有具有日本古代传统特征的"五佑卫门浴盆"、传说洗后能多子多孙的"子宝之泉"、一年四季都可使用的"25米温泉游泳池"等等。在这里，游客与自然山水融为一体，春天赏花，冬天看雪，夜里可以仰望漫天星斗，侧耳聆听原野天籁之声，让游人流连忘返。

天城庄有28座温泉浴池，包括室内浴池、户外浴池、洞穴浴池、瀑布浴池、温泉泳池（冬天开放）、岩石浴池、家庭浴池等等，若要一一尝试，必须用上几天时间。

1. The panoramic view of the Japanese architecture in the vilage

2. The open-air bath named "Kawara no Yu". There are totally five pools in dfferent temperatures, which can be chosen according to individual body condition. Being a part of the valley, it is an imaginary experience to bathe in the water mist sprayed from Odaru Waterfall.

3. "Dousojin no Yu", the hot spring bath with shallow stream, the ideal bath for admiring the beautiful natural scenery

1. 和式风格建筑物外景，坐落于小村庄中

2. 露天浴池"河原之泉"。置身于溪谷，融入"大瀑"激扬起的水雾之中，如幻如诗。"河原之泉"共有5个浴池，每个浴池水温各异，可根据身体状况选择不同的浴池

3. "道祖神之泉"，水浅，适于躺着欣赏自然美景的温泉浴池

4. Women's open-air bath, which has a panoramic view of the blooming sakura in February every year

4. 女士专用露天浴池，从这里可以饱览每年2月河津盛开的樱花

5. The indoor pool for men called "Iwa Buro", which was built of the local rock of Amagi

6. View from half-open bath pool for women, in which you can overlook the red maples, green trees and mountains, bringing you a pleasant mood

5. 男士专用室内大浴池之"岩浴池"，采用天城本地岩石建造而成

6. 女士专用半露天浴池，名为"野天浴池"。浸于其中，可远眺周边红枫、绿树、青山，使人心情舒畅

7. The indoor bath pool for women

8. The indoor bath pool for men equipped with "Kodakara no Yu", "Herbal Medicine Bath", and the like

7. 馆内女士专用大浴池

8. 男士专用室内大浴池，设置有"子宝之泉"、"草药浴池"等

9

9. The open-air bath pool in the special guest room

10. The balcony of the guest room. Most of the building materials come from the local wood in Amagi, creating a natural sense.

9. 特别客房内的露天温泉浴池

10. 客房阳台。客房大部分建筑材料为天城本地树木，给人一种自然感

11

11. View to the outside from the balcony of the guest room

12. The restaurant named "Wakatake", with comfortable seats, capacious dinning space and the open kitchen which offers various kinds of freshly cooked cuisine, including the diversified Japanese buffet breakfast

13. The banquet hall which can accommodate 120 persons, equipped with a stage and the professional acoustic equipments

11. 透过客房阳台观赏室外美景

12. 名为"若竹"的餐厅有舒适的座椅和宽敞的就餐空间。开放式厨房提供新鲜的各式料理。早餐为品种多样的和式自助餐

13. 可同时容纳120人的宴会大厅，设有舞台及专业音响设备

Mizuniwa no Hatago Sumiyoshikan
水庭笼住吉馆

Location: 1149-1, Kona, Izunokuni City, Shizuoka 410-2201, Japan

Spa Name: Kona Onsen

Area: 2,549.53 m^2

Quantity of Guest Room: 20

Quantity of Spa Facility: 4

Other Facilities: private open-air bath, rentable open-air bath

地址：日本静冈县伊豆之国市古奈1149-1

温泉名：伊豆长冈古奈温泉

占地面积：2549.53平方米

客房数：20

温泉浴池数：4

其他设施：私人专用露天浴池、包租露天浴室

Sitting on the riverside of Kano River in central Izu Peninsula, Mizuniwa no Hatago Sumiyoshikan is a Japanese hotel permeated with the natural beauty of Izu. The various guest rooms with unique design offer a comfortable and pleasant experience. The hot spring inside the hotel is called Kona Onsen, whose history can be traced back to the Kamakura Period (1185~1333). Kona Onsen boasts spa baths in various styles such as the large stone open-air baths and the chartered open-air baths, "Keika no Yu".

The room "Fugatei" facing Kano River, commanding a panoramic view of the ever-changing scenery in all seasons, is a traditional Japanese guest room where guests can overlook the rolling mountains of Hakone. The annexe named "Retoro", exquisite and elegant, encompasses the traditional sense of beauty and the modern comfort and conciseness. The guest rooms inside the annexe are equipped with the chartered open-air spa baths and Japanese beds.

The food here is the typical Japanese Kaiseki Ryori, elaborately cooked with fishes from Suruga Bay and the fresh seasonal vegetables, which are full of delicious flavour and abundant nutriment.

水庭笼住吉温泉旅馆位于日本伊豆半岛中部狩野河河畔，是一个渗透着伊豆自然美的日式风格旅馆。具有独特设计主题的各个客房，让游客的旅行更加舒心满意。旅馆内温泉源泉名为"古奈温泉"。"古奈温泉"历史久远，在日本的镰仓时代（1185－1333年）就开始使用。古奈温泉设有以岩石砌成的露天大浴池以及包租露天浴室"桂花之泉"等各种风格的温泉浴池。

面朝狩野河"风雅亭"的是可以眺望箱根连绵群山，饱览四季更迭美景的传统日式客房。名为"别邸Retoro"的别馆，风雅别致，既显出传统美感，又具有现代的舒适简洁。

别馆的客房内有专用露天温泉浴池、和式睡床等。水庭笼住吉温泉旅馆的料理为日式怀石料理，食料是从骏河湾捕到的鲜鱼以及各种时令蔬菜，经过精心烹制，味道鲜美、营养丰富。

1. The panoramic view of "Fugatei"

2. The night scene of the tranquil and serene Japanese garden

3. The small bridge over the flowing stream in the garden

4. The corridor with reception spaces

1. 风雅亭外观

2. 庭园夜景——宁静祥和的日式风格庭园

3. 庭园的小桥流水

4. 带有会客空间的走廊

5. The vestibule in the warm and gentle light

5. 散发着温暖柔和灯光的迎客玄关

3

4

6. The private open-air bath. In August, the blooming osmanthus blossom fills the room with fragrance.

7. The special guest room with wooden and ceramic bath tubs to let guests experience a different kind of bathing pleasure

6. 租赁露天私用温泉浴室，到了八月，庭园中的桂花盛开，更是醇香满屋

7. 特别客房中设有专用木制和陶制浴盆，感受不同沐浴乐趣

8

8. The guest room in Japanese style with the exclusive open-air bath pool

9. The guest room combining the unique modern and traditional design style

10. The bed in the guest room

8. 带专用露天浴池的和式风格客房

9. 融合现代和传统风格，设计独特的客房

10. 客房睡床

Nishiizu Toi Onsen Hotel Miyabi

西伊豆土肥温泉酒店——雅

Location: 241 Nishiizu Toi, Izu, Shizuoka Prefecture, 410 - 3301, Japan

Spa Name: Nishiizu Toi Onsen

Area: 4,890 m²

Quantity of Guest Room: 17 with an accomodation capacity of 90 persons

Quantity of Spa Facilities: 4

Other Facilities: private open-air bath

地址：日本静冈县伊豆市小土肥241

温泉名：伊豆土肥温泉

面积：4890平方米

客房数：17 / 容纳人数 90人

温泉浴池数：4

其他设施：专用露天浴池

Nestled in western Izu Peninsula, Nishiizu Toi Onsen Hotel was built with the advantage of the nearby Toi Gold Mine, a well-known scenic spot. Facing the White Sand Beach formed by the scouring waves in Suruga Bay, the hotel stands next to the cliff "Koyibito Misaki", famous for the hanging Love Clock. Guests can listen to the vivid rhythm of the waves in all the guestrooms that designed in modern Japanese style. In the setting sunlight, the combination of the golden sky and the sea looks as glare as an oil painting.

Being observed in the distance, the black granite and the red granite add radiance and beauty to each other. Guests can admire the stunning night scene of the small coastal town. They can also taste the banquet cuisine cooked with the seafood from Izu Peninsula, which has won the most public praise. The staff from South Korea and China can provide all-round services.

位于伊豆半岛西部的土肥温泉旅馆依傍日本著名观光景点土肥金山而建。土肥温泉旅馆周边是以挂有爱之钟而闻名的恋人岬，正面是骏河湾波浪冲刷而成的白沙海滨。全部客房洋溢着现代和式风格，游人可以侧耳倾听海浪波涛的韵律。夕阳西下，放眼望去，那染成了金黄色的天空和海洋宛如一幅油画。

在土肥温泉大浴场，黑御影石和红御影石交相辉映。从露天浴池的露台可以欣赏到美丽的海滨小镇夜景。在客房内游客可以品尝到伊豆半岛口碑一流的海鲜宴会料理。这里聘有韩国及中国籍员工，可以让游客获得全方位的服务。

1. The facade of the Japanese spa hotel

2. The vestibule and the entrance

3. The exclusive open-air stone bath pool in the guest room, installed on 16-square-metre terrace

1. 面向大海的日式温泉旅馆

2. 玄关、入口

3. 客房专用露天岩石浴池，设在约16平方米的露台上

4. The ocean-view open-air pool

4. 海景露天浴池

5. The guest room with the high-grade cypress bath, where you can enjoy the stunning scene of the sky and the broad Toi sea

6. Teahouse

5. 带有高级桧木浴池的客房，可悠闲地享受海阔天空的美景

6. 茶室

7. Standard ocean-view guest room in Japanese style, approx.16 square metres (about 10 tatami mats), together with the wide corridor

8. The guest room with the open-air bath pool

9. The after-bath lounge

7. 标准和式海景客房，约16平方米（10张榻榻米）及宽走廊（日语称"广缘"）

8. 带有露天浴池的客房

9. 浴后休息沙龙

Toi Hotel Sankaitei

山海亭土肥温泉酒店

Location: 324 Nishiizu Toi, Izu, Shizuoka Prefecture, 410–3301, Japan

Spa Name: Izu Toi Onsen

Area: 1,430.89 m²

Quantity of Guest Room: 40

Quantity of Spa Bath: 7

Other Facilities: private open-air bath, chartered spa

地址：日本静冈县伊豆市土肥324

温泉名：伊豆土肥温泉

面积：1430.89平方米

客房数：40

温泉浴池数：7

其他设施：私用露天浴池、包租温泉

Located in western Izu Peninsula, Toi Hotel Sankaitei boasts the ideal location of famous scenic spots commanding a panoramic view of the gloriours scenenry of Toi in sunset glow. Guests can listen to the vivid tempo and melody of tide and spray and enjoy the stunning view of the setting sun while lying in the observatory open-air bath pool. The great flavour and interests of nature make guests intoxicated in endless enjoyment.

There are 21 guest rooms with exclusive open-air bath pools, and all are made of high-grade cypress. The modern Japanese style is fully demonstrated by the water-cycled natural spa and elegant and cosy guest rooms.

The observatory chartered open-air bath pool named "Yakata Takarabune" was made with the technology of the ship constructor. There are seven different kinds of spas such as the outdoor bath in the garden, the observatory open-air bath (Mabu Hot Spring), which are all available to guests.

Guests can leisurely enjoy fresh marine fish from the Toi Sea and vegetables prepared for breakfast and dinner in the guest rooms.

山海亭土肥温泉酒店位于伊豆半岛西部，那里最适宜观赏伊豆土肥落日余晖。在开阔的露天浴池里，游人可以边泡温泉，边聆听浪花合奏，同时还能欣赏到夕阳西下的美景。这种无拘束的大自然情趣给游人带来无尽的享受。

山海亭土肥温泉酒店内有专用露天浴池的客房共21间，全部采用高级桧木材料打造，提供循环式纯天然温泉服务。客房装饰典雅、舒适，充分体现了现代日式风格。

酒店内的包租露天浴池，名为"屋形宝船"，采用造船师的工艺技术制作。另有庭园野天浴池、展望露天浴池（Mabu汤）等共7种风格各异的温泉，可随意自由享用。

游客可以在客室内悠闲地享用早餐和晚餐，所用食材为新鲜蔬菜和当天从土肥海域捕获的鲜鱼。

1. The rooftop open-air spa "Mabu Hot Spring" – exclusive for men

2. The lobby

3. The high-grade-cypress-made open-air bath pool in the shape of a wine goblet

4. All guest rooms in the new building "Ajisaitei" are equipped with open-air bath pool

1. 屋顶露天温泉，男士用"Mabu汤"

2. 大厅

3. 樽型高级桧木露天浴池

4. 新馆"味彩亭"的每间客房内都设有露天浴池

5. The observatory open-air bath pool "Yakata Takarabune", built with exquisite workmanship. It is so pleased to overlook Suruga Bay while bathing in it.

5. 露天浴池"屋形宝船"，工艺精湛，入浴时在此展望骏河湾，无限惬意

Ochiairou Murakami

落合楼村上温泉旅馆

Location: 1887-1 Yujima, Izu, Shizuoka Prefecture, 410-3206, Japan

Spa Name: Izu Yujima Onsen

Quantity of Guest Room: 15

Quantity of Spa Bath: 4

Other Facilities: private open-air spa, rentable spa, foot spa

地址：日本静冈县伊豆市汤岛1887-1

温泉名：伊豆汤岛温泉

客房数：15

温泉浴池数：4

其他设施：私用露天浴池、包租温泉、足汤

As a perfect example of incorporating Japanese and Western building style, Ochiairou Murakami has been written in National Cultural Heritage lists. It is a rare traditional wooden hotel. In Meiji 14 (1881 AD), Yamaoka Tesshū (1836-88AD), the Warrior, suggested renaming the hotel as Ochiairou, when he overlooked Kano River, the intersection of Hontani River and Neko River, and then it was known as "Ochiairou". Since the beginning of Meiji period, numerous literators have stayed at the onsen ryokan during their trips. The vestibule and the annexe named "Minuntei" were added continuously from Showa 8 to Showa 12 (1933-1938 AD). In the Japanese traditional academy-style guest rooms with antique beauty, the niche is installed next to the curio box shelf and the partition board door, which were both made with the mosaic craft, the same technique used in the design of the window lattice. They were both handmade by carpenters without using any iron nails or any adhesive, demonstrating the amazing craftsmanship.

Guest rooms can be classified into two categories: the ones facing the garden and the ones in the face of the river, which are heated by wood fire in winter and cooled by fresh air and clear stream gushing from Mount Amagi in summer, making guests fully relaxed in mind and body.

落合楼村上温泉旅馆是日本传统建筑与现代风格完美结合的典范，已被列入日本国家文化遗产。这是一座罕见的传统木造旅馆。明治十四年，武士"山冈铁舟"来到此地，遥望本谷河与猫越河交汇的狩野川，提议将原建筑更名为"落合楼"，"落合楼"因此而得名。自明治时期开业以来，无数文人墨客来此温泉旅馆下榻。玄关和别馆"眠云亭"是昭和八年至十二年增建。古色古香的客房采用日本传统建筑"书院式"，壁龛旁是多宝格架和隔扇门，精美隔扇门及窗棂均采用拼花工艺制作，不使用铁钉及黏合剂，完全凭木匠手工组装而成，精湛的工艺技巧令人叹为观止。

客房分为"庭园风景"以及"凭窗临河"两种。冬天，室内用木薪烤火取暖；夏日，从天城山涌出的清流与清新的空气，让人体验身心放松之旅。

1. Ochiairou Murakami, surrounded by sakura in early spring

2. The front vestibule of the onsen ryokan in the ancient fine wooden structure, which was listed as the National Cultural Heritage

3. The chartered spa, in which you can admire the murmuring stream and river, breathe the fresh air and enjoy the cosy sunshine

4. The open-air bath pool, named "Tengu no Yu", comprising the stone bath pool and the brook-front bath pool

1. 早春的落合楼村上樱花环绕

2. 正面玄关,被列为国家级文化遗产的古老木结构温泉旅馆

3. 浸泡在包租温泉里,天城山溪流潺潺、空气清爽、风光宜人

4. 露天浴池"天狗之汤",由岩石浴池和望溪浴池构成

5. The façade

5. 外观

6. The Japanese banquet hall, adorned with delicate red sandalwood carving, simple and mellow

7. The reading room

8. The guest room named "Tsubaki" with Ashiyu (foot bath), where you can have a foot spa while observing the beautiful Kano River

6. 用精美的紫檀木雕装饰的日式宴会大厅，古朴醇厚

7. 阅览室

8. 设有足汤（足浴温泉）的客房"椿"，倚着窗边泡足汤，狩野川从眼前流过

9. The luxury guest room "Koubai" in the annexe "Minuntei"

10. Bedroom

9. 别馆 "眠云亭" 的豪华客房 "红梅"

10. 客房卧室

Hotel Izukyu

伊豆急酒店

Location: 2732 – 7 Shirahama, Shimoda, Shizuoka Prefecture, 415 – 8512, Japan

Area: 3,600 m²

Capacity: 416 persons

Quantity of Spa Facilities: 4

Quantity of Guest Room: 117

地址：日本静冈县下田市白滨2732－7

温泉名：伊豆白滨温泉

面积：约3600平方米

容纳人数：416人

温泉浴池数：4

客房数：117

Sited in the south of Izu Pennisula, Hotel Izukyu was built facing the snow-like Shirahama Coast, one of the eight wonderful scenes of Izu, which can be seen from both the reception hall and the guest rooms. Shirahama Onsen is unique in that it was excavated from its own properties, comprising two hot springs: Ryujin Onsen and Rendaiji Onsen, with their respective water quality and function; the former one has the special efficacy in treating arteriosclerosis, dermatosis and helps recovering from diseases, whereas the latter is exceptional in its function for relieving neurodynia, myalgia and arthralgia. The Department of General Affairs and the Marketing Department of the hotel were nominated as one of the "One Hundred Top Japanese Hotels" in the year of 2010.

伊豆急温泉酒店位于日本伊豆半岛南端，正面为伊豆八景之一的白滨海岸。伊豆急温泉酒店临海而建，从会客大厅和客房就能眺望到如雪一般的白沙海滨。白滨温泉的特别之处在于它的源泉是自内部挖掘而成，名为"龙神之泉"和"莲台寺温泉"。两泉最大特点是泉质不一，功能各异。"龙神之泉"对动脉硬化、皮肤疾患和病后恢复等具有特殊功效。"莲台寺温泉"主要对神经、肌肉、关节疼痛等具有特别功效。酒店的综合部和企划部入选了2010年度"日本百家优质酒店"之一。

1. Night view. It offers you a totally different feeling to walk and play leisurely in the courtyard with lights everywhere.

2. The façade – the outdoor swimming pool in front of the hotel, where you can enjoy yourself swimming, while admiring the blue sky, the blue sea and Shirahama Coast

3-4. The open-air spa named "Banza Furo" with the spring resource from Ryujin Onsen, having the special efficacy in treating arteriosclerosis, dermatosis and helping recovering from diseases

1. 外观夜景——灯火映照时分在庭园散步、游玩，别有一番情趣

2. 外观——酒店前是露天泳池，夏季里您可以边游泳嬉戏，边享受面前的蓝天大海、阳光沙滩

3–4. 露天温泉，名为"磐座浴池"。"龙神之泉"对动脉硬化、皮肤疾患、病后恢复等具有特别功效

5

5. The indoor spa bath – the clean and spacious bathing pool with the spring resource from Rendaiji Onsen, having the special efficacy in relieving neurodynia, myalgia and arthralgia and the function of bubble massage

5. 室内温泉大浴池。整洁宽敞，源泉为"莲台寺温泉"，对神经，肌肉，关节疼痛等等具有特别功效。还具有气泡按摩功能

3

4

6

6. The indoor spa bath — the clean and spacious bathing pool with the spring from Rendaiji Onsen, having the special efficacy in relieving neurodynia, myalgia and arthralgia

7. The reception room in the hall, bright and transparent, warm and cosy

8. The garden restaurant named "Kagayaki" offering buffet breakfast

6. 室内温泉大浴池。整洁宽敞，源泉为"莲台寺温泉"，对神经、肌肉、关节疼痛等等具有特别功效

7. 大厅接待室明亮通透，阳光洒入带来暖洋洋的舒心惬意

8. 庭园餐厅"绚烂"提供品种丰富的自助早餐

11

9. The Japanese-style guest room

10. The luxury suite with the best location commanding a panoramic view of the picturesque view of the spacious sea, the snow-like Shirahama Coast at all seasons

11. The European-style guest room with simple but grand arrangement

9. 日式风格的客房

10. 日式豪华套房，占据最佳的眺望角度，透过窗户尽览宽广无垠的大海、四季如雪的海滨白沙

11. 欧式风格的客房，布置简单大方

Hoshinoya Karuizawa
虹夕诺雅–轻井泽温泉酒店

Location: Hoshino, Karuizawa-machi, Nagano 389-0194, Japan

Spa Name: Hoshino Onsen

Area: 42,055 m²

Quantity of Guest Room: 77

Quantity of Spa Facilities: 2

地址：日本长野县轻井泽町星野

温泉名：星野温泉

占地面积：42055平方米

客房数：77

温泉浴池数：2

"What would our life and the surrounding scenery be like if we reserve Japanese traditions while keep moving to modernization?" Hoshinoya Karuizawa has created another world of Japan under this ideal, which gave birth to the beautiful sights of "Countryside in the Valley".

All the guestrooms in Hoshinoya Karuizawa were built as single houses, thus a village was formed by the lanes connecting the houses. Those houses are now dotted along the creek, which runs through a small valley with the lush forests. In front of Mount Asama, Karuizawa is among the top few summer resorts in Japan. With cool climate in summer, Karuizuwa is located on the 1000-metre-high plateau area, where the Emperor met his lifetime mate by chance. In accordance with the different locations and construction procedures, the guestrooms here can be classified into Mizunami (the riverside villa), Yamaroji (the hillside villa) and Niwaroji (the garden villa). As an original Japanese holiday resort hotel showcasing the culture of Japan, Hoshinoya Karuizawa has been used to convey the Japanese culture to the world. Focusing on offering guests the chance of experiencing the great flavour and interest of Japanese hotel and the world top-class service with the value of protecting

the environment and constituting a sound ecosystem, Hoshinoya Karuizawa endows guests with the superior Japanese service abreast of the times.

"如果能在保留日本传统特色的同时走向现代化，大家的生活与周围的风景将是个什么模样呢？"为了寻找答案，虹夕诺雅开始了新的计划。后来，在这一构思之下诞生了"山谷村落"的人间美景。

虹夕诺雅–轻井泽温泉酒店的客房均为单门独户型，连接它们的小巷形成了一片村落。别墅式客房散落在森林包围的山谷中。轻井泽是日本屈指可数避暑胜地之一，地处海拔约1000米的高原地带，背靠浅间山，夏季气候凉爽，日本天皇就是在这里邂逅了一生的伴侣。虹夕诺雅中所有的客房，按照地理位置和建造方法的不同分为水波房、山路房和院庭房。"虹夕诺雅"作为日本起源的"和"文化日式度假酒店，把"和"文化传达给世界。酒店以"体验日本旅馆的趣味、提供世界顶级服务"为中心，坚持保护资源环境、生态系统的价值观，展现当地风情文化，使嘉宾享受符合时代的顶级日式服务。

1. The overview of the village in summer

2. The panoramic view of the resort in the setting sun. Focusing on the theme of the resort hotel with the elements of Japanese villages, Hoshinoya Karuizawa is a typical sustainable hotel with a resource self-sufficiency rate of about 70%

3. The Meditation Bath – the meditative space to soothe and relax the soul, comprising Hikari-no-heya with bright light and the dark "Yami-no-heya" with the drifting faint sound

4. The interior of the Hoshinoya spa, providing the traditional Japanese Shiatsu massage, the oil treatment, to the original Reiki massage

1. 夏天村落全景

2. 村落夕阳全景。"虹夕诺雅-轻井泽"是典型的环保酒店，能源资源的自给自足率达到70%

3. 虹夕诺雅冥想温泉，由两部分空间构成，释放光线的"光房"（Hikari-no-heya）和飘荡微弱声响的"暗房"（Yami-no-heya）

4. 虹夕诺雅SPA室内，提供肌肤营养护理和指压按摩，原创按摩等服务内容

5. The panoramic view of Tombo-no-yu

5. 星野温泉（Tombo-no-yu）全景

1. 夏天村落全景

2. 村落夕阳全景。"虹夕诺雅-轻井泽"是典型的环保酒店，能源资源的自给自足率达到70%

3. 虹夕诺雅冥想温泉，由两部分空间构成，释放光线的"光房"（Hikari-no-heya）和飘荡微弱声响的"暗房"（Yami-no-heya）

4. 虹夕诺雅SPA室内，提供肌肤营养护理和指压按摩，原创按摩等服务内容

6. The exclusive spa room in HOSHINOYA Karuizawa

7. The full view of the Japanese cuisine "Kasuke" cooked with the typical Japanese cooking ideas of "Five Flavours, Five Methods and Five Senses", the healthy Japanese courses that embrace tradition and feature the local delicacies and fresh ingredients from Karuizawa

6. 虹夕诺雅SPA专用房

7. 日本料理嘉助全景。秉承日本料理的精髓思想"五味、五法、五感"。在向传统致敬的同时，以轻井泽独有的乡村时鲜食材为主

10

8. The living room of the hillside villas, which were built as single isolated villas leaning against the mountains; the room on the tall platform facing the birds and the forests

9. The living room in the garden villa; the self-contained room within the bardian garden

10. The overview of the guest room on the hillside; spacious room with complete facilities

8. 虹夕诺雅山路客房起居室，依山面向野鸟森林的高台上的房间

9. 虹夕诺雅庭院客房起居室，特有个性的庭园，是独门独户型的客房

10. 虹夕诺雅山路客房全景，宽敞的空间里配置丰富

Hotel Hananoi

花乃井温泉酒店

Location: 399 – 1 Kanznjicho, West District, Hamamatsu, Shizuoka Prefecture, 431-1209, Japan

Spa Name: Kanznjicho Onsen

Area: 2,775 m²

Quantity of Guest Room: 14

Quantity of Spa Bath: 4

Other Facilities: private open-air bath, rentable bathroom

地址：日本静冈县滨松市西区馆山寺町399－1

温泉名：馆山寺温泉

面积：2775平方米

客房数：14

温泉浴池数：4

其他设施：私人专用露天浴池、包租浴室

Hotel Hananoi enjoys the ideal location on the fields of fortune of Kanznjicho, which was firstly built in 810 AD, as early as 1,200 years ago. Kanznjicho Onsen was the great work by the designer who had successfully designed Hoshinoya Karuizawa and Izumi Hoshimo Hotel. With the central concept of harmony between the colours of gold and cyan-blue, and the unity of Japanese and Western styles, the interior design of high-class Liukiuu tatami in the gentle candle light, leads tourists immediately to a tranquil and cosy world.

Since its opening in 1958 (Shouwa 33), Kanznjicho Onsen has been widely used in that the appropriate temperature of salt solution has significant curative effect to neuralgia and rheumatalgia. Kanznjicho Onsen incorporates Koto no yu, the open-air spa, in which guests can feel the pleasant cool breeze, and Hananoyu including a large number of indoor spas such as the ones with massage bathtubs and saunas. The open outdoor spa and meditation spa, which are exclusive to men and women at different time intervals, offer guests differed spa experiences.

The guestrooms are located in the space from the fourth to the sixth floor all with a panoramic view of the Hamanako Lake, comprising the double-bed room with the cyclic annular floor, presidential room, Hananoi modern room with Liukiuu tatami, Hananoi premiere modern room, suite room with outdoor bath and the room for the disabled.

花乃井温泉酒店依托馆山寺这方宝地而建，馆山寺始建于公元810年，距今已有1200年。馆山寺温泉是曾经设计虹夕诺雅–轻井泽和星之野京都旅馆的设计师亲自主持设计。设计以"日式与欧式"，"黄金也丹青"的和谐统一为主题。室内采用高级硫球榻榻米，附以柔和的蜡烛灯光，让游人很快进入恬静悠然的世界。

该温泉于1958年（昭和三十三年）开放，因食盐水适宜的热度对神经痛及风湿痛有很好疗效而被人们广泛利用。馆山寺温泉"花乃井"，具有让游人感受凉爽微风的全开放式露天温泉——"湖都温泉"以及配备按摩浴缸、桑拿等多种室内温泉服务的"华的温泉"两种。开放的露天温泉和冥想温泉，采用男女交替使用的方式让游客的温泉体验更多样。

客房设在能够看到滨名湖景色的5~7层。特别设计环状式地板的双人房、总统双人房、琉球榻榻米式的花乃井现代房、花乃井高级现代房、带有露天温泉的特别房以及残疾人房六种规格。

1. The atrium, the living environment design with the background of the natural environment, with the central design elements of "Universal Design"

2. The interior of the open-air spa "Koto no yu", the bath made of high-grade cypress

3. The exterior of the open-air spa "Koto no yu", where you feel sank in the tranquility with the breeze of Hamanako Lake passing by

4. The Japanese guest room incorporating the Western style – the interior space features the high-grade Liukiuu tatami, with the central idea of the harmony and unity between the colours of gold and cyan-blue and the combination of Japanese and Western style

5. The exclusive indoor bath made from superior cypress

1. 中庭注重以自然环境衬托生活环境，以"无障碍设计"（Universal Design）为设计要素

2. 露天温泉"湖都之汤"的室内高级桧木浴池

3. 露天温泉"湖都之汤"的室外，滨名湖微风轻拂，感觉自己被静谧包围

4. 与西洋风格融合的日式客房采用高级琉球塌塌米，设计以"日式与欧式"，"黄金也丹青"的和谐统一为主题

5. 专用高级桧木浴池

6. The façade of the hotel

6. 酒店外观

1. 中庭注重以自然环境衬托生活环境，以"无障碍设计"（Universal Design）为设计要素

2. 露天温泉"湖都之汤"的室内高级桧木浴池

3. 露天温泉"湖都之汤"的室外，滨名湖微风轻拂，感觉自己被静谧包围

4. 与西洋风格融合的日式客房采用高级琉球塌塌米，设计以"日式与欧式"，"黄金也丹青"的和谐统一为主题

4

5

IDUMISOU

星野IZUMISOU温泉旅馆

Location: 2 – 21 Okahiro-cho, Ito-city, Shizuoka Prefecture, 414-0016, Japan

Spa Name: Ito Onsen

Area: 5,264 m²

Number of Guest Room: 34

Spa Facilities: private open-air bath, chartered bath, foot spa

地址：日本静冈县伊东市冈广町2 – 21

温泉名：伊东温泉

面积：5264平方米

客房数：34

其他设施：专用露天浴池、包租浴室、足浴

IDUMISOU lies in Izu Peninsula on the south of Tokyo. Ito spa comprises more than 750 hot springs with such an abundant stream volume that it can spout over 32,000 litre spring water per minute, which makes it one of the top famous spa resorts in Japan. With the concept of "Paradise with Unique Tranquility", IDUMISOU spa hotel boasts not only various spas indoor and outdoor, but also spas for family use and the foot spa, making guests feel at home.

Founded in 1912, IDUMISOU is available to guests all year round. From the spa bath and the open-air bath made of ancient cypress to the graceful 6,600 square metres' Japanese gardens in different styles, and even the fresh flowers eager to bloom all seasons, whose beautiful images are reflected in the spring water with vertical flow as well as the cosy and amiable natural oxygen bar, all these make guests feel so cheerful in mind and body that they forget all fatigue. They can also enjoy the seafood cuisine of Izu.

星野IZUMISOU温泉旅馆位于东京南部的伊豆半岛。伊东温泉共有源泉750多处，泉涌量极为丰富，每分钟可喷涌至少3.2万升的泉水，是日本全国屈指可数的温泉度假胜地。星野IZUMISOU温泉旅馆设计以"娴静天地"为主题，拥有室内室外各式温泉、足浴，让游客有一种宾至如归的感觉。

星野IZUMISOU温泉旅馆始建于1912年，全年无休。该温泉不仅设有温泉泳池和古代丝柏木制作的露天浴池，还有造型优美、风格各异的6600平方米日式庭园，一年四季竞相开放的鲜花映衬在垂流式的温泉中，形成舒适宜人的天然氧吧，这一切让人疲惫尽消、身心愉悦。在这里，游人还可以品尝到伊豆特有的海鲜料理。

1. Founded in 1912, IZUMISOU spa hotel is designed with the concept of "Paradise with Unique Tranquility".

2. The night scene of the swimming pool with hot water, which is available all year round, forming a delightful contrast finely with the Japanese garden

3. The open-air bath for women encapsulated by trees. The drinking water of the waterfall and the stone bath comes from the headspring.

4. The indoor bath for women where you can admire the courtyard with the open-air bath and feel the natural oxygen bar permeating with the water mist splashed from the waterfall

1. 于1912年开业的温泉旅馆，以"娴静天地"为设计主题

2. 源泉温水游泳池夜景，游泳池与花木丛生的日式庭园相映成趣

3. 被树木包围的女士露天岩石浴池，流下的瀑布和岩石浴池内的饮用水，全部来自源泉

4. 女士室内浴池，置身桧木大浴池，欣赏露天浴池庭园，垂落瀑布飞溅的水雾弥漫着天然氧吧的气息

5. The Japanese courtyard – the 6,600 square metres' courtyard in which the flowers are eager to bloom all seasons

5. 面积约6600平方米的日式庭园，春夏秋冬鲜花竞放

3

4

6. Night scene of the independent Bettei guest room circled by the beautiful courtyard

7. The exclusive open-air cypress bath pool

8. The Bettei guest room located in the two-storey building equipped with the half-open bath built from Ito stone and the open-air bath made of ancient cypress

6. 别邸客房，庭园美景环绕的客房夜景

7. 专用露天桧木浴池

8. 别邸客房——独立的"竹泉"客房为二层和式小楼，内部配有伊豆山石砌成的半露天浴池及由古代丝柏木制作的露天浴池

9. The elegant guest room in Japanese style, in which you can smell the fragrance of cypress, the original building material of the indoor bathroom

10. The Japanese restaurant "Zuiun", against the background of the picturesque traditional Japanese garden

11. The Japanese-style guest room with study room, dressing room, porch and eaves gallery, each having its unique layout

9. 格调高雅的和式房间，专用浴室采用高级桧木，房间内桧木飘香

10. 日式餐厅"瑞云"，传统日式庭园如餐厅的背景画一般

11. 和式房间，设有书房、化妆间、门廊、檐廊等，每套客房布局各异

ANGINE

星野ANGINE温泉旅馆

Location: 5-12 Nagisa-cho, Ito, Shizuoka Prefecture, 414-0023, Japan

Spa Name: Ito Onsen

Area: 6,031 m²

Quantity of Guest Room: 38

Quantity of Spa Bath: 4

Other Facilities: indoor exclusive spa

地址：日本静冈县伊东市渚町5-12

温泉名：伊东温泉

面积：6,031平方米

客房数：38

温泉浴池数：4

其他温泉：客房专用温泉

Facing the sea in the beauteous Ito Peninsula, Hoshino ANGINE spa hotel was designed by Japanese star architect Rie Azuma. Combining the traditional and modern design style, the hotel offers private spaces far away from metropolis life, endowing them with spiritual tranquility. With the name "ANGINE", derived from the name of the adviser on foreign affairs for "Tokugawa Leyasu", British navigator "William Adams". It features the colourful space such as the modern ocean-view Japanese-style guest room, the guest room with private spa and typical Japanese courtyard, surrounded by the stunning attractions and entertainments, such as Jyogasaki Coast, Izo Park, the fireworks in memory of the navigator "William Adams", the beautiful flowers and scenery as well as the open-air spa bath and the indoor spa bath called "Skincare Hot Spring", a series of spaces which are available to guests all the time during their stay.

The spas here all have their unique water qualities such as the alkalescence one and the hypotonicity one. It is interesting that the different water qualities are particularly effective in skincare, alleviating neuralgia, rheumatism, arthralgia and paralysis motorica.

星野ANGINE温泉旅馆坐落于美丽的伊东湾，面朝大海，由日本著名建筑师东理惠氏设计，是一个远离喧嚣都市、融合传统和现代风格，注重设计的温泉旅馆。旅馆名字"ANGINE"源于担任德川家康外交顾问、英国航海家三浦按针（William Adams）的名字。这里为游人准备了面向大海的现代日式客房、带露天温泉的客房、纯传统式日本庭院等多彩的空间，周边有城崎海岸、伊豆公园，还有纪念航海家三浦按针的烟花活动以及伊东的鲜花美景。旅馆设有露天温泉浴池、名为"美肤之泉"的室内温泉浴池等，游客居住期间可以随时体验。

星野ANGINE温泉旅馆的温泉具有不同的泉质，不同的泉质具有不同的疗效（弱碱性、低张性等），对美肤、神经疼痛、风湿关节痛、运动麻痹等都有明显功效。

1. The façade of the spa hotel facing the sea with its unique shape

2. The open-air bath for women, whose different water qualities have different effects. It is so interesting to admire the stunning landscape in the Japanese courtyard.

3. The evening view of the Japanese-style guest room

4. The reading room on the way between the guest room and spa pool, where there are abundant books and picture albums about Ito

1. 面朝大海、造型独特的温泉酒店外观

2. 女士专用露天浴池。ANGINE温泉旅馆的各温泉因为不同的泉质分别具有不同的疗效。沐浴后即可在日式庭园中游赏，别有情趣

3. 纯日式独立客房（"离"）的傍晚风景

4. 浴池与客房途中设置了阅览室。泡完温泉后可在此放松。这里准备了丰富的关于伊东的书籍、画册等

5

5. The 2,640 square metres' Japanese Courtyard with picturesque trees and flowers, making guests intoxicated in the harmonious atmosphere of gardens and buildings

5. 日式庭园，面积2640平方米，四季花木竞相开放。从窗中看园景，园林与建筑相映成趣

3

4

8

6. The guest room "Hanare" with open-air spa

7. The restaurant "Gosutan" provides all-private-room service, enjoying the delicacies in the peaceful and cosy environment

8. The guest room mixed Japanese and Western style with a view of Ito Peninsula

6. "离"，带有露天温泉的独立客房

7. 全包间餐厅"吴须丹"，在沉静、舒适的环境中，享受品尝美味佳肴

8. 日式与欧式融合的客房，在追求现代时尚美感的客房中能看到伊东湾

Horai

星野蓬莱温泉旅馆

Location: 750－6 Mount Ito, Atami, Shizuoka Prefecture, 413－0002, Japan

Spa Name: Atami Onsen

Area: 4,900 m²

Quantity of Guest Room: 16

Quantity of Spa Bath: 3

地址：日本静冈县热海市伊豆山750－6

温泉名：热海温泉

面积：4900平方米

客房数：16

温泉浴池数：3

Located on Mount Ito, Atami which has a panoramic view of Sagami Bay, Horai is founded in Kaei 2 (1849 AD), the end of the Edo period, a famous hotel with a history of about 160 years. The name "Horai" refers to a place on the eastern sea, where an ageless immortal lived. The guests can experience the unique paradise-like space. Various seasonal cuisine are served here with the pure flavour of the ingredients. All the guest rooms are both in traditional Japanese aroma such as the Japanese-style guest room on a high platform, the one with a courtyard, creating the atmosphere of Japanese traditional culture, as well as the building with the large open-air spa in Eastern traditional building style. Located in Shizuoka Prefecture near Tokyo, it is surrounded by attractions including Izu Mount Shrine with the unique culture of Japan.

As one of the Three Great Ancient Onsens in Japan, Izu Onsen is called Holy Spa by the faith followers, as it is next to Izu Mount Shrine. The spring quality of the vitoriol spring is particularly effective in curing empyrosis and incise lesion.

There are two open-air spas in Horai, "Hashiri-yu", which means running hot water and "Kogoi-no-yu", which is designed by Kengo Kuma, a famous architect.

星野蓬莱温泉旅馆位于伊豆山上，相模湾的美景尽收眼底。创建于江户时代末期嘉永二年（1849年），是一所具有160年历史的传统老字号旅馆。蓬莱，传说位于东方的海上，在那里居住着长生不老的神仙。入住这里如同进入仙境般的空间。这里根据不同季节准备不同特色的料理，最大的特点是充分体现食物本身的美味。旅馆的客房具有日本传统韵味，设有高台式、庭院式的日式房间，散发着日本传统文化气息。旅馆设有露天大温泉与传统的东屋风式的汤殿。旅馆处于距离东京不远的静冈县，周边的伊豆山神社等景观能令游客充分感受日本特有的舒适。

伊豆山温泉是日本三大古泉之一，因其流经伊豆山神社而为人们信仰，被称作是神的温泉，属于硫酸盐泉，对烧伤、割伤等疗效显著。旅馆两处露天浴池分别为喷涌而出的"走之汤"与隈研吾设计的"古古比之汤"，游客可在温泉中观赏大海，度过一段美好时光。

1. The guest room with the tranquil Japanese style, where the attractive views inside and outside create different feelings

2. The façade of the superior Japanese spa hotel located in the peaceful Izu Mount

3. The open-air spa "Hashiri-Yu"

4. Kogoi-No-Yu designed by Kengo Kuma Architects

1. "离"，幽静的日式独立客房，室内及室外庭园美景营造出情趣各异的氛围

2. 纯日式高级温泉旅馆外观，坐落于伊豆山中幽静之处

3. 露天浴池"走之汤"

4. 隈研吾设计的"古古比之汤"

5. The lobby, in which the considerate services with Japanese cultural spirits attracts guests all over the world

5. 大厅，饱含日本文化精髓的温馨服务，迎接各界贵宾来此下榻

3

4

6. The teahouse-like guest room permeated with the flavour of the traditional culture of Japan

7. One of the guest rooms in the traditional Japanese style

8. The luxury Japanese-style guest room decorated with ornaments implying the spirits of Horai

6. 茶室风格客房，散发日本传统文化气息

7. 具有日本传统特色、韵味的客房

8. 豪华日式客房，室内的装饰用品件件都体现了旅馆的内涵

Kishoan

星野贵祥庵温泉旅馆

Location: 1-31-1, Matsumoto City, Nagano Prefecture, 390-0303, Japan

Spa Name: Asama Onsen

Area: 3,464 m²

Quantity of Guest Room: 26

Quantity of Spa Bath: 13

Other Facilities: private open-air spa

地址：日本长野县松本市浅间温泉1-31-1

温泉名：浅间温泉

面积：3,464平方米

房间数：26

温泉浴池数：13

其他设施：专属露天浴池

Located in eastern Matsumoto City in central Nagano, Kishoan was designed by famous modern architect Takao Habuka & S.E.N Architects Associates. The hot springs in Kishoan are called "Taka No Yu" and the well-known "Zenkouji Onsen", which incorporate 82 kinds of spring stream, providing different spa experiences. All the guest rooms and the spas enjoy the individual design with the cob wall and the Japanese paper as the keynote, taking the open-air spa "Kiten" as an example, whose natural wooden material expresses a sense of beauty, matched with round windows. Whether the entrance hall or the corridor or even the guest rooms, are designed by the designer with unique themes, whilst the guest rooms, moreover, seem quietly elegant in the embellished light, featuring the brisk colour and tone.

Combining the elegant façade and the contemporary design, Hoshino Kishoan Spa endows the whole hotel with a grand style. Kishoan Onsen is exclusive to men and women at different time intervals in the morning and in the evening. The simple alkaline water of the 13 characteristic spas including the radiant heated bath and the open-air spa, has significant effects to neuralgia, gastroenterology, gynecological diseases, dermatitis and diabetes mellitus.

Guests can taste different Shinshu Kaiseki, the local food of Shinshu with the highest-grade wine, changed with seasons.

星野贵祥庵温泉旅馆位于长野县中部的松本市东部，是日本当代著名建筑设计师羽深隆雄的作品。星野贵祥庵温泉源泉为"鹰之汤"和著名的"善光寺温泉"，两个温泉殿有82种泉水涌入，拥有自家温泉的贵祥庵温泉旅馆为游客提供不同的温泉体验。设计者以土墙和日本和纸为基调对每间客房和温泉浴池进行个性化设计，如"贵天"露天浴池，采用天然珍稀木材建造，配以圆窗显出"和"的美感。从入口大厅、走廊到房间，建筑师坚持每个空间都有不同的艺术主题。客房更是以清新的色调为主，在灯光的点缀下，透出如诗般的淡雅，融合了日本传统文化与现代气息。

星野贵祥庵温泉以风雅的外观与现代感完美结合使旅馆整体焕发出设计大师的风范。贵祥庵源泉早晚男女交替使用。水质是碱性纯净水，有大温泉辐射热温浴、露天温泉等13种特色温泉，对治疗神经痛、肠胃病、妇科病、皮肤病、糖尿病有很大的功效。

旅馆根据不同季节，为客人准备信州怀石料理——特有的食物配上极品葡萄酒。

1. The waterfall fountain in the patio

2. The façade of the spa resort hotel; the traditional elements designed in modern style by the famous Takao Habuka & S.E.N Architects Associates

3. The indoor spa named "Kiten", made with the rare natural wood – an indoor bath with extreme openness, produced by the patio stretching out to the forum

1. 中庭瀑布喷泉

2. 外观——将传统以现代方式呈现的贵祥庵度假温泉旅馆，是日本当代著名的建筑设计师羽深隆雄的作品

3. "贵天"室内大浴池采用天然珍稀木材建造，向外庭延展的天井，造就了开放感十足的室内浴场

4. The detail of the indoor communal spa named "Kiten", producing a poetic elegance with the ornament of the lighting

4. 局部特写，灯光点缀下，透出诗般淡雅

5. The open-air spa named "Kiten", in which the design of the round window expresses a kind of Japanese beauty

6. The open-air bath in the guestroom, made of natural black granite

7. The corridor to the guestroom, and each part was designed with its unique artistic theme by the designer

5. "贵天"露天浴池，圆窗显出"和"的美感

6. 客房的露天浴池，由天然黑御影石制成

7. 通往客房的走廊，每个空间都有不同的艺术主题

8. The panoramic view of the guestroom with open-air spa

9. The guestroom with open-air spa, in which the gentle light and the spring scenery of the garden can be seen through the lenient windows on the terrace

10. The guestroom with open-air spa

8. 客房露天浴池全景

9. 带有露天温泉的客房。透过露台宽大的落地窗，柔和的光线和满园的春色映入眼帘

10.带露天温泉的客房

Sun Hatoya Hotel

太阳鸠屋温泉酒店

Location: 572 – 12, Yukawa, Ito-City, Shizuoka Prefecture, 414 – 0002, Japan

Spa Name: Ito Onsen

Area: 16,141 m²

Quantity of Guest Room: 191

Quantity of Spa Bath: 8

地址：日本静冈县伊东市汤川坚岩572 – 12

温泉名：伊东温泉

面积：16141平方米

客房数：191

温泉浴池数：8

The coastal location of Sun Hatoya in central Izu Peninsula commands a panoramic view of the stunning seascape for each room. In the submarine hot spring, the only one in Japan, guests can enjoy the spa bath. With the central theme of "Ancient Monuments—Submarine Adventure", the indoor swimming-wear-required hot spring zone includes Exploration Zone, Mystery Fall Zone, Stream Swimming Pool Zone and Children's Swimming Pool Zone, among which you can admire the roaming fish while bathing in Fish Bathing Pool. Being a unique scenery line of the hotel, the aquarium-like spa endows guests with the pleasant sensation in enjoying the cosy spa and the underwater world. The spa bath with historical remains and the outdoor swimming pool nearby are bathing places where guests need to wear their swimming wears. The indoor spa bath with historical remains is available in all seasons. The name of the submarine spa derives from the natural flow gushed out of the 1,000-metre-deep submarine hot spring, which is rarely seen in Japan and the spas in all rooms come from the same spring resource .

Monthly different dishes are served here, beside performance in the banquet hall with a capacity of 1,000 persons. There are dancing hall with a capacity of 800 people, the game rooms and the karaoke.

太阳鸠屋温泉酒店位于伊豆半岛中部，所有的房间都可以看到美丽的海景。这是日本唯一一座海底温泉，在海底温泉，游人可以穿泳衣享受温泉洗浴。以"古代遗迹、海底冒险"为设计主题的室内泳装温泉区分为"发掘区"、"探险瀑布（Mystery fall）"、流水游泳池、儿童游泳池等。其中，鱼浴池（鱼风吕）可以一边享受温泉一边观赏很多热带鱼在水槽里漫游。这个像海族馆般的温泉，可谓是酒店特有的一道风景，让客人同时享受舒心温泉和精彩海底世界。旁边的古代遗迹温泉浴池和室外泳池是泳装温泉区。室内的古代遗迹温泉浴池，不受气候影响，四季都可以利用。"海底温泉"之所以得名，是因为使用了海底1000米自喷的海底温泉，这在日本国内也是罕见的。酒店所有房间的洗浴也都用同一源泉的温泉。

该酒店每月推出不同菜肴接待游客，晚宴设在可容纳1000人的剧场宴会厅，享用美餐之余还可以欣赏表演秀。酒店还设有可容纳800人的交谊舞厅和游戏厅、大宴会厅、卡拉OK等设施。

1. The night scene of the appearance of the hotel standing on Ito coast

2. The submarine spa

3. The submarine spa

4. The ocean-view restaurant named "Daigyouen", which features the fresh seafood cuisine

1. 外观夜景

2. "海底温泉"

3. "海底温泉"

4. 以新鲜海鲜料理为主的大鱼苑餐厅，可眺望大海

5

5. The layout of the submarine spa

5. "海底温泉"平面设计图

6. The banquet hall, enjoy daily night performances by the well-known troupe with actors from across the world. There are 626 seats on the ground floor and 202 on the first floor.

7. The banquet hall with performance stage

8. The banquet hall with performance stage

6. 剧场宴会厅每晚安排海内外著名演出团的表演。1层有626个坐席，2层有202个坐席

7. 剧场宴会厅

8. 剧场宴会厅

Hotel Hatoya
鸠屋温泉酒店

Location: 1391 Oka, Ito, Shizuoka Prefecture, 414–0055, Japan

Spa Name: Ito Onsen

Area: 11,379 m²

Quantity of Guest Room: 162

Quantity of Bath: 2

地址：日本静冈县伊东市冈1391

温泉名：伊东温泉

面积：11379平方米

客房数：162

温泉浴池数：2

Located in Oka, Ito Peninsula, Hotel Hatoya and Sun Hatoya are two hotels of the same brand. The most distinguishing feature of the hotel is that guests here can play the submarine adventure game using the facilities in the submarine spa of Sun Hatoya Hotel, the chain hotel, enjoying the great pleasure in the spa and the underwater world just like the guests in Sun Hatoya Hotel.

The five large spa pools with local natural hot spring are highly praised by spa experts for its appropriate water temperature for human body and the skin care. The guest can not only enjoy the spa bath, but also oversee the stunning scene of Ito in the guestrooms permeated with elegant atmosphere. It is really a very good choice for holiday.

There are over 70 kinds of choices including seafood and local food for buffet dinner.

鸠屋温泉酒店与太阳鸠屋温泉酒店是连锁姊妹酒店，位于伊豆半岛的伊东市，鸠屋温泉酒店的最大特色是客人可以同时体验太阳鸠屋温泉酒店海底温泉的"海底冒险"游戏，享受到与太阳鸠屋温泉酒店客人同等的招待。

酒店设有5个引用自家天然源泉的大浴池。这里的温泉水温适宜，不仅使人体感觉舒适，也有助于保护皮肤中的有益成分，为此受到温泉专家的高度赞誉。客房设计洋溢着幽雅情趣，在客房中不仅能享用温泉沐浴，还可以眺望伊豆风光，是一家非常适合度假的温泉酒店。

酒店提供自助晚餐，海鲜、山中野味等70个种类可自由选择。

1. The façade

2. The indoor spa pool for women

3. The indoor spa pool for women

4. The indoor cypress-made bath pool

1. 外观

2. 女士室内大浴池

3. 女士室内大浴池

4. 室内桧木大浴池

5

5. The reception hall

5. 接待大厅

3.

4.

6. The indoor large communal bath for men

7. The meeting room named "Kuzyu" with an area of 145 square metres and a capacity of 60 persons

8. The meeting room named "Orange" with an area of 111 square metres and a capacity of 90 persons

6. 男士室内大浴池

7. 会议室"九重"，面积145平方米，可容纳60人

8. 会议室"橘"，面积111平方米，可容纳90人

7

8

9. The banquet hall with performance stage, with an area of 850 square metres and a capacity of 510 persons on the ground floor and 152 persons on the first floor

10. The entertainment room with table-tennis

11. The Japanese-style guestroom

9. 剧场宴会厅，面积850平方米，1层可容纳510人，2层150人

10. 乒乓娱乐休息室

11. 日式客房

10

11

Sora Togetsusou Kinryu

宙Sora渡月庄金龙

Location: 3455 Shuizenji, Izu-shi, Shizuoka Prefecture, 410-2416, Japan

Spa Name: Shuizenji Onsen

Area: 49,500 m^2

Quantity of Guest Room: 29

Quantity of Spa Bath: 4 indoor spas, 4 outdoor spas

Other Facilities: private open-air spa, rentable outdoor spa

地址：日本静冈县伊豆市修善市3455

温泉名：修善寺温泉

面积：49500平方米

客房数：29

温泉浴池数：室内温泉4、露天温泉4

其他设施：专用露天温泉浴池、包租露天浴池

Shuizenji Onsen Zone is a spa street along Katsuragawa. It is said that the hot spring was made by the eminent monk "Kobo Daishi" with its withered stick in order to cure the old mother's sickness whose son work hard all day looking after her, which made the monk deeply moved. The original hot spring is the Shuizenji Onsen today. It was in this advantaged location that the Japanese spa hotel with the modern concept was designed. Sora Togetsusou Kigru boasts a private courtyard with an area of 49,500m², encompassing 45 to 49 temples of the nearby tourist attraction "Keikoku 88".

Sora Togetsusou Kinryu focused on its concept with the use of light, whose transparency creates a unique space feeling. The most outstanding spa is the only-one "Lights Open-Air Bath" in Japan designed by the famous interior designer, Tsujimura Hisanobu. The unique spa was highlighted in almost all magazines in Japan and had won the Excellent Award in the NASHOP Lighting Awards. Bathing in the lighted-glass-made pool seems like travelling in the universe.

The unique design can also be found in the private dinning room "Kumo-no-niwa", whose walls are all embedded with spotlights, making guests feel like dinning in a lightened box. However, the food in this fashionable restaurant are exquisite traditional Kaiseki (the advanced Japanese feast). The ideal location on the hillside of Mount Shuizenji enables guests to admire the stunning sceneries around while enjoying the cuisine.

修善寺温泉区是沿着"桂川"延展的一条温泉街。据传，公元807年，高僧弘法大师在此地被一位不辞劳苦照顾病弱母亲的孝子感动，以枯杖在桂川击出泉眼，流出的温泉水治好了孝子的母亲，此泉眼即是今日修善寺的独钴汤温泉池。宙SORA渡月庄金龙依其得天独厚的地理位置，设计出充满现代概念的日式温泉旅馆。宙SORA渡月庄金龙拥有49500平方米专属庭园。修善寺周边著名的"桂谷八十八所"景点中的45~49所都在该旅馆的庭园内。

"宙"在设计上以"光"来呈现其主题。大量运用灯光透射营造不同的空间感。其中最特别的是日本唯一的"光之露天浴池（风吕）"，是日本著名室内装潢设计师辻村久信的作品。它几乎在日本各大杂志的温泉特辑上亮过相，曾获得松下电工株式会社授予的优质服务部门"优秀奖"。浴池采用发光的玻璃灯箱打造，浸泡其中如同浮在"宙"（宇宙）中一般。

包间餐厅"云庭"的设计上也充分体现了设计主题。"云庭"墙身镶满了射灯，就像在发光的箱子内用膳。客人身处充满现代气息的餐厅，享用的仍是考究的日式传统"怀石料理"（日式高级宴席）。餐厅位于修善山腰，面朝山野，可以边用餐，边欣赏四周景色。

1. The façade of restaurant "Kumo-no-niwa", locating on the hillside of Mount Shuizenji enables guests to admire the stunning sceneries around while enjoying the cuisine

2. The garden in autumn – the traditional Japanese garden with an area of 49,500 square metres

3. The vestibule in summer

4. The pathway in the bamboo forest

1. 包间餐厅"云庭"外观，餐厅位于修善山腰，面朝山野，可以边用餐，边欣赏四时景色

2. 拥有49500平方米的传统日式庭园秋景

3. 玄关夏景

4. 竹林小径

5

5. The façade of the hotel, surrounded by mountains; the sunlight transmitted through the woods leads guests to the paradise

5. 位于山的怀抱之中，从树林里透射出的阳光把游人带到童话般世界

竹林の小径

6

7

6. The surroundings of "Dokko No Yu"

7. The open-air spa made of natural stone

8. The vestibule in autumn

6. 独钴汤周边环境

7. 天然岩石打造的露天温泉

8. 玄关秋景

9. The interior space of "Open-Air Spa of Sunset" which means admiring the beautiful setting sun while bathing

10. The exterior space of "Open-Air Spa of Sunset"

11. "Open-Air Bath of Lights" all made of light boxes, a lightened open-air spa at night

9. "晚霞"，露天浴池室内，取意"一边泡汤一边看夕阳"

10. "晚霞"，露天浴池室外

11. "光之露天浴池"，用灯箱打造，每到夜晚就变成发光的露天浴池

12

12. One of the guest rooms facing the mountains and the fields with red leaves, comprising a guest room with an area of 20 square metres, 9.9-square-metre corridor and the indoor spa

13. One of the guest rooms facing the mountains and the fields with green trees, comprising a guest room with an area of 20 square metres long and 9.9-square-metere corridor and the indoor spa

14. The corridor of the restaurant "Kumo-no-niwa" whose walls are all embedded with spotlights

12. 客房，可观赏满山的红叶，由12.5帖榻榻米（20平方米）+宽走廊（9.9平方米）和室内温泉构成

13. 客房，可观赏翠绿山林，由12.5帖榻榻米（20平方米）+宽走廊（9.9平方米）和室内温泉构成

14. 餐厅"云庭"走廊，墙身全部都镶满了射灯

Atami Sakuraya Ryokan

热海樱屋旅馆

Location: 9–11 Higashikaigancho, Atami-shi, Shizuoka Prefecture, 413-0012, Japan

Spa Name: Atami Onsen

Area: approx. 1,983 m²

Quantity of Guest Room: 21

Quantity of Spa: 4

地址:日本静冈县热海市东海岸町9－11

温泉名：热海温泉

面积：约1983平方米

客房数：21

温泉浴池数：4

It is only five minutes walk to Atami Sakuraya Ryokan from Atami Railway Station and three minutes walk from the Atami Beach. Hided in the forests with the surrounding green trees in the central area of Atami, it enjoys the ideal location in the hinterland, the scenic spot of Atami where you can see the first plum blossom and the Atami onsen had won praise from the first general Tokugawa leyasu.

With the basic service concept of "the only lucky chance in life", the hotel makes guests fully relaxed here. The open-air spas comprise the superior cypress bath and the stone bath made of the natural Izu rock. Various spa baths are available to men and women at different time intervals.

The guest rooms are in two types: "Touzankaku" (Ocean Front Guest Rooms) and "Sihoukaku" which are in various styles such as the Japanese style, the Japanese and Western style and the double deck. All guest rooms are equipped with spa baths with their own characteristics and elegant tastes. The most outstanding one is the highly-praised Japanese and western guestroom with a large bed and the Kotatsu on the first floor of Touzankaku, in which guests can have a panoramic view of the surrounding mountains and rivers through the open window.

The cuisine features the luxurious feast with seasonal seafood from Suruga Bay, which can be enjoyed in the guest rooms.

热海樱屋旅馆距日本国铁"热海站"步行仅需要5分钟，距热海海滨步行3分钟。热海樱屋旅馆位于热海中心地带，隐藏在绿荫环绕的山林里，地处"日本第一早梅"名胜景区腹地。热海温泉曾深受日本第一代将军德川家康（ 1543 — 1616年）的喜爱。

该旅馆把"有生之年的唯一机缘"作为服务宗旨，让游人能够尽情地放纵身心。露天温泉为高级桧木浴池和天然伊豆石的岩浴池，早晚男女交替制，可以享用风格各异的温泉浴池。

客房分为"陶山阁"（海景客房）和"漆宝阁"两个类型。内设多种房型：日式、日欧融合式、楼中楼等。客房设有温泉浴池，室内风格雅趣相宜，各具特色。特别是陶山阁二楼的日欧融合式客房，既配有宽大的床铺、落脚式暖炉，房间窗体开阔，可纵览山水风光，深受好评。

旅馆料理选用骏河湾的时令海鲜，客人可以选择在客房内享用。

1. The façade of the quiet and beautiful hotel

2. The "Kaisei No Yu" with baths of the horizontal type

3. The guest room equipped with the exclusive open-air bath pool, connected with the courtyard

4. The spa pool named "Kaisei no Yu"

5. The spa pool named "Tensei no Yu" with sauna

1. 静美的旅馆外观

2. 大浴池"海星之汤"，设有卧式浴

3. 客房配有专用露天浴池，与庭园相连

4. 大浴池"海星之汤"

5. 大浴池"天星之汤"，设有桑拿浴

6. The hall in ancient style

6. 复古式大厅

4

5

7. The guest room

8. The standard guest room with overview of Sagami Bay through the large window

9. The guest room of mixed Japanese and European style

7. 客房

8. 标准客房窗体开阔，一览相模湾海景

9. 日式与欧式融合的客房

8

9

Yagyu-No-Sho
竹庭–柳生庄温泉旅馆

Location: 1116-6, Shuzenji, Izu-shi, Shizuoka, 410-2416, Japan

Spa Name: Shuzenji Onsen

Area: 10,000 m²

Quantity of Guest Room: 15

Quantity of Spa Bath: 4

地址：日本静冈县伊豆市修善寺1116－6

温泉名：修善寺温泉

面积：10000平方米

客房数：15

温泉浴池数：4

Yagyu-No-Sho is a luxurious high-end traditional hotel in the well-known Shuzenji Onsen area. Built in 1970 as the annexe of the high-class Tokyo Ryotei "Yagyu-no-Sho", it was designed in the form of the bamboo court, creating a natural sense of beauty. The main building was built in the traditional style. There are 15 Japanese guest rooms and 2 independent apartments. The annexe, the independent guest rooms named "Matsu-no-o" and "Ume-no-o", are perfectly matched with the natural environment such as green bamboos, the mountain forest, the pool, the light and shadow. All the space from the vestibule to the guest room is permeated with the sense of Japanese Zen. Guest rooms were designed in above ten Japanese styles such as the one in the style of the academy of science, the teahouse style and the farmhouse style, implying perfect combination of the traditional Japanese culture, modern culture and humanity.

The hypotonic low alkaline hot spring (PH8.7) has the efficacy of softening the cuticle, and 65mg moisturising elements per kilogramme in the water make it "Spring of Skin Care".

In the original intention of making guests taste the gourmet in the natural environment of Ryotei, the owner built the private hot spring Ryotei. With 30-year-cooking experience, the head chef "Sibayama Takashi" is good at cooking Izu local ingredients such as Izu beef, Izu tricholoma matsutake and Ise lobster in different methods, offering guests various kinds of delicious food.

豪华传统旅馆竹庭－柳生庄温泉旅馆位于著名的修善寺温泉区。1970 年，作为东京高档料亭"柳生庄"的别邸而建，在设计上采用竹庭的形式，实现了一种自然和谐美感。主楼采用日本丝柏，以传统数寄风格制造。柳生庄温泉旅馆有和式客房15间、独栋寓所2间。别邸中的独立客房"松生"和"梅生"与室外葱茏的绿竹山林、池塘光影等自然环境完美融合。独特的日本禅味渗透到从玄关到房间的每一个角落。客房风格各异，荟萃日式风情，有书院风、茶室风、田舍家风等十几种客房，可以感受到传统的日本文化与现代文化、人性化的完美结合。

竹庭－柳生庄温泉旅馆的水质为低张性、弱碱性温泉（酸碱值pH8.7）具有软化角质层效果，再加上每千克65毫克的偏硅酸成分，使其成为保湿效果良好的"美肌"温泉。

在大自然中品味"隐秘的温泉料亭美食"是馆主的初衷。总厨柴山崇志入行近30年，擅长使用伊豆当地时令食材，如伊豆牛、伊豆松茸、伊势龙虾等材料，配合不同的烹调方式让游人尽享柳生庄美味。

1. The vestibule in autumn

2. The façade of the top exquisite Ryotei spa hotel located in the deep tranquil area of Shuzenji

3. The exterior bath pool of the open-air spa "Musashi-no-yu"

1. 玄关秋景

2. 坐落在修善寺深处娴静的、顶级温泉旅馆

3. "武藏"露天浴池的室外浴池（外汤）

4. The vestibule in summer, built with Tamajyari

4. 玄关夏景及门前玉砂利

6

5. The independent guest room in the annexe named Matsuo

6. The exclusive bath pool in the guest room "Yuuzuki"

5. 别邸独立客房"松生"，让住客在房内也能欣赏室外的庭园竹林

6. 客房"夕月"专用浴池

7. The open-air bath "Ichiju"

8. The indoor bath pool of the open-air spa "Musashi-no-yu"

9. The ceramic bath pool of "Fish Pavilion"

7. 客室"一树"露天浴池

8. "武藏"露天浴池的室内浴池（内汤）

9. "游鱼亭"陶制浴池

10

10. The guest room "Yuuzuki"

11. The flower arrangement – the interior decoration in autumn

12. The Japanese guest room "Hatsukari", with semi-open spa pool

10. 客房"夕月"

11. 秋季室内装饰，插花

12. "初雁"和式客室，带有半露天浴池

ASABA Ryokan
浅羽温泉旅馆

Location: 3450 – 1, Shuzenji, Izu-shi, Shizuoka Prefecture, Japan

Spa Name: Shuzenji Onsen

Quantity of Guest Room: 17

Capacity: 50 persons

Quantity of Spa Bath: 3

Other Facilities: chartered spa 2

地址：日本静冈县伊豆市修善寺3450 – 1

温泉名：修善寺温泉

客房数：17

容纳人数：50

温泉浴池数：3

其他设施：包租温泉浴池2个

Located in Monzenmachi, Shuzenji, Izu-shi, Shizuoka Prefecture, ASABA Ryokan was found in 1489 with a history of more than 520 years, when the ancestor Asaba Yakurouyukitada, who was the general following Chan master Ryukeihansho, the founder of the mountain, came to guard a temple. It was those ancestors who created the oldest spa hotel in Shuzenji Temple. At the end of Meiji, under the protection of Asaba, the seventh owner of the hotel, Yasuemon removed the Gekkeiden Noh Stage here from Fukagawa Tommiokahachimangu, which made ASABA Ryokan a well-known high-ranking spa hotel with Noh performance in Japan. With a garden area of 1,980 square metres, ASABA Ryokan is surrounded by green hills and blue water. Attracted by the spa water quality and abundant spring flow, a large number of guests come here to see Noh and Munchausen, performed here numerous times in a year. Surrounded by the bamboo forest, the interior decoration and the exterior building are both in the traditional Japanese style, overlooking the large and broad pond. The open-air spa, hidden in the lush green bamboo forest, with the birds' twitter and fragrance of flowers, the whirling and murmuring sound of the bamboo forest and the fresh air. Moreover, there are also Koya-Maki Baths, Baths for Ladies, Baths for Gentlemen and Family Baths.

浅羽温泉旅馆位于静冈县伊豆市修善寺的门前町，该旅馆的历史要追溯到520年前，1489年（延德元年），先祖浅羽弥久郎幸忠作为修禅寺曹洞宗开山之祖——隆溪繁诏禅师的随将来到修善寺，由此产生了最古老温泉旅馆。明治后期，第7代馆主浅羽保佑卫门从东京深川富冈八幡宫的能舞台"月桂殿"迁移到了这里。浅羽温泉旅馆四周青山绿水环绕，庭园面积有1980平方米。从此，浅羽成为日本有名的上演"能"的高级温泉旅馆。每年有数回"能"和"狂言"的舞台表演。很多游人被修善寺温泉的水质和丰富的流量所吸引。该旅馆为竹林所包围，俯瞰宽大的池塘，内部装饰和外部建筑均为传统日式风格，隐藏在翠绿竹林之中的露天温泉，伴随鸟语花香、竹林婆娑、空气清新，十分受游人欢迎。此外，还有高野槇浴池、妇人和殿方浴池、家族浴池等也可以享用。

1. The panoramic view of the famous spa hotel with a long history of several hundred years

2. The panoramic night scene of the hotel

3. The vestibule

1. 有数百年历史的著名温泉旅馆全景

2. 夜晚全景

3. 玄关

4. The Gekkeiden Noh Stage, the Noh drama stage built with authentic pine

4. 能舞台"月桂殿"，正宗柏木建的能乐戏台

5. The Gekkeiden Noh Stage, a Noh drama theatre opposite the hotel, removed from Fukagawa Tomioka Hachimangu Shrine, Tokyo

6. The outdoor pool Koyamaki-wood bath

7. The artistic lounge & salon

5. 能舞台"月桂殿"，客栈对面是一座能乐剧院，是从东京深川富冈八幡宫移筑过来的

6. 高野槙露天浴池

7. 艺术休息沙龙

8. The spa for ladies   8. 女士浴池

9. The guest room   9. 客房 "萌葱"

10. The guest room   10. 客房

HANAFUBUKI Ryokan

花吹雪温泉旅馆

Location: 1041 Yawata, Ito-shi, Shizuoka Prefecture, 413-0232, Japan

Spa Name: Izukogen Spa

Area: approx. 13,000 m²

Capacity: 45 persons

Quantity of Guest Room: 17

Quantity of Spa Facilities: 7

地址：日本静冈县伊东市八幡野1041

温泉名：伊豆高原城柯崎温泉

面积：约13000平方米

容纳人数：45

客房数：17

温泉浴池数：7

Located in Yawata, Ito-shi, Shizuoka Prefecture along the Fuji Mount, Hakone and the Kinosaki coastline of Izu Natioanl Park, HANAFUBUKI Ryokan was designed with the concept of "Onsen Forest", while the most outstanding design was the spa in the jungle. When entering, guests are led by the gate to a green world of a large area of forest and the open-air spa also surrounded by jungle. The spas are made of natural wood such as cypress and Kuromoji. Taking Kuromoji-Yu as an example, it was made of the local tree called Kuromoji, which produces a light smell of aroma when sank in the hot spring, thus becoming guests' favourite. Whether in the indoor or the outdoor spa, guests can breathe the fresh air from the forest and are set in a mood surpassing that of the immortals in the quiet courtyard with the moss floor, the bug buzz and the birds' song. They can simply go to any of the seven Onsen, get in, lock the entrance, and enjoy Onsen bathing freely, with families or with friends. All onsens here are available 24 hours.

HANAFUBUKI Ryokan features the Kaiseki cuisine with "Sakura" theme favoured by tourists, which can offer guests the most original local gourmet of Izu Plateau.

The most popular HANAFUBUKI breakfast menu hasn't been changed for three years and has won many awards.

花吹雪温泉旅馆位于静冈县伊东市八幡野。沿着富士山、箱根、伊豆国立公园的城崎海岸线，就可以到达伊豆高原城柯崎温泉的"花吹雪"日式温泉旅馆。该旅馆以"温泉森林"为设计理念，旨在打造密林中的温泉，从玄关入口"森之门"开始便把游客带入一个绿色葱茏的世界。花吹雪温泉旅馆拥有一大片森林，其露天温泉浴池都被密林包围着。温泉浴池的材质均为柏木、黑文字香木等天然木材，例如，黑文字汤温泉浴池用的就是当地一种名为"黑文字"的树木。这种木头浸泡在温泉中可以散发出一丝清淡的木香，因而深受游人喜爱。游客在旅馆的各处温泉浴池均能感受到来自森林的清新空气，深深庭院、青苔铺地，伴随着鸟叫虫鸣，让游客体验到赛过神仙的愉快享受。旅馆有7种温泉供客人任意挑选，锁上门就可以与家人享受专属温泉空间。全部温泉24小时开放。

花吹雪温泉旅馆还有广受游客推崇的"樱怀石料理"，与传统的"京怀石"相比别具特色，为客人们提供最正宗的伊豆高原当地美食。备受欢迎的花吹雪早餐菜单三年来保持不变，并获奖无数。

1. The vestibule entrance "Gate of the Forest", the beginning of the cosy holiday

2. The independent guest room of Hakuo Villa hidden in the forest

3. Kuromoji-Yu spa including the indoor and the outdoor ones made from the local tree called Kuromoji, which produces a light smell of aroma when sank in the hot spring

4. The half-open-air spa made from Izu stone and cypress, with fresh breeze from the forest coming through the opened window

1. 玄关入口"森之门",隐匿惬意的假日从这里开始

2. "白翁别墅"——隐匿于森林中的独立客房

3. 黑文字汤温泉浴池。香木"黑文字"是当地一种树的名字。这种木头浸泡在温泉中可以散发出一丝清淡的木香

4. 半露天浴池,使用伊豆石、柏树而造成。森林中的清新之风透过敞开的窗户徐徐吹来

5. The night scene of the hotel

5. 旅馆夜景

1. The vestibule entrance "Gate of the Forest", the beginning of the cosy holiday

2. The independent guest room of Hakuo Villa hidden in the forest

3. Kuromoji-Yu spa including the indoor and the outdoor ones made from the local tree called Kuromoji, which produces a light smell of aroma when sank in the hot spring

4. The half-open-air spa made from Izu stone and cypress, with fresh breeze from the forest coming through the opened window

1. 玄关入口"森之门",隐匿惬意的假日从这里开始

2. "白翁别墅"——隐匿于森林中的独立客房

3. 黑文字汤温泉浴池。香木"黑文字"是当地一种树的名字。这种木头浸泡在温泉中可以散发出一丝清淡的木香

4. 半露天浴池,使用伊豆石、柏树而造成。森林中的清新之风透过敞开的窗户徐徐吹来

6. The wooden outdoor balcony of Hakuo Villa

7. The entrance of Hina-No-Yu

8. Oribe-Yu decorated with Oribe-Yaki

6. "白翁别墅"的木板阳台

7. "雏之汤"的入口

8. "织部温泉"采用织部陶器的特制装饰

7

8

9. Shino-No-Yu decorated with Shino-Yaki

10. Kuromoji-Yu spa interior

11. The bath tub of Hina-No-Yu made from the clean Izu stone and cypress

9. "志野之温泉"使用日本陶器装饰

10. 黑文字汤的室内浴池

11. "鄙之汤"的浴槽采用有整洁感的伊豆石和柏树制成

10

11

12

13

12. Hina-No-Yu Spa built in 2006, where you can enjoy the simple and beautiful atmosphere far away from the metropolis

13. The spacious and tranquil interior space of Hakuo Villa

12. "鄙之汤"于2006年建造,散发远离都市的质朴之美

13. "白翁别墅"室内宽敞、沉静

14. The Japanese conversation room with a capacity of 4 to 5 people

15. The guest room, where you can catch sight of the garden on the balcony

16. The small Japanese guest room of Hushi Villa in the teahouse style

14. 和式风格的会谈室可供4~5人使用

15. 最受客人欢迎的客房"赤",透过阳台可一览室外庭园美景

16. "西行梦樱"小型和室,设计采用茶室风格

Isawa View Hotel

石和VIEW饭店

Location: Nakajima 1607, Isawa-cho, Fuefuki City, Yamanashi Prefecture, 406-0024, Japan

Spa Name: Isawa Onsen

Area: 4,750 m²

Quantity of Guest Room: 50

Quantity of Spa Bath: 4

Other Facilities: private open-air spa, chartered spa

地址：日本山梨县笛吹市石和町川中岛1607

温泉名：石和温泉

面积：4,750平方米

客房数：50

温泉浴池数：4

其他温泉：专属露天浴池、包租浴室

It takes only one and a half hour from Tokyo to Isawa View Hotel by trolley bus, which is located in Kawanakajima, Isawa-cho, Fuefuki City, Yamanashi Prefecture. Isawa Onsen Town is the largest hot springs in Yamanashi Prefecture. One kilometre away is the Sakuragi standing in a line on each side of the river, and then the hotel can be seen at the end of the "Sakuragi Onsen Street". With elegant facilities, the hotel has long been pursuing the refined Japanese sensibility and sincerity. The guest rooms are in pure Japanese style and are created to provide a calm ambience, in which you can understand the spirit of the Japanese culture and time flowing slowly in harmony. Mt.Fuji can be viewed from the south-oriented rooms on a fine day. In the open-air spas blessed with unique amorous feelings, the name "Keiryu" endows the gentlmen's spa pool with more calmness and "Seiryu" brings more rustic atmosphere to ladies' spa pool. The guests can warm their bodies to the core while feeling the breezes of the seasons in the clear air of the rotenburo of which the hotel is very proud.

The stature of Ikkyu san, one of the most famous three monks in Japanese history, half lying down at the entrance, welcomes guests from across the world. While entering the hall, guests will see the design with the theme of "Interweaving Light and Water" immediately. Time seems stop while staying with the swimming fish silently.

The experienced chef cooks Kaiseki Ryori food with abundant seasonal ingredients of Yamanashi.

石和VIEW温泉饭店位于山梨县笛吹市石和町川中岛。从东京乘电车到石和VIEW温泉饭店只需1.5小时。石和温泉乡也是山梨县最大的温泉地。距该馆大约1千米处可以看到沿河流两岸一字排开的樱树，走过这条"樱木温泉大道"，就可以看到这家饭店。饭店设施格调高雅，秉承"日式风"的洗练、感性与真诚。客房全部采用纯和风设计，稳重、超然，同时体现出日本文化内涵，让客人体验到缓缓流逝的"日式风"时光。天气晴朗时从南侧客房可眺望富士山风光。男用露天温泉名为"溪流"，稳重深沉；女用浴场名为"清流"，充满野趣。酒店引以为傲的露天温泉为游客带来四季吹拂的微风、清净的空气和温暖舒适的水。

门前半卧着"一休"师傅的石像（日本僧侣史上最有名的三位和尚之一），微笑着迎接八方来客。步入大厅会让人立刻感受到以"光水交织"为主题的设计风格。在大厅旁的池塘静静观赏锦鲤鱼（本地特产）游动时的优雅姿态会让人忘记时光的流逝。

会席料理由经验丰富的厨师选用山梨地区当季多种食材精心烹制而成。

1. The façade detail

2. The façade of the hotel

3. The fish pond next to the hall

4. The gentlmen's open-air spa

1. 外观一角

2. 饭店外观

3. 大厅旁的锦鲤鱼池塘

4. 男士露天温泉

5. The vestibule and the statue of Ikkyu san, greeting guests from across the world

6. The hall with the remarkable electrolier lamp in the shape of a gamp and the adjacent pond full of fancy carps

5. 玄关，门前"一休"的石像迎接宾客

6. 大厅，大伞形状设计的大吊灯引人注目

3

4

9

7. Gentlmen's spa, "Keiryu", the spacious space that boasts abundant hot water and a calm atmosphere, equipped with massage bathtubs and bubble bathtubs

8. Ladies' spa, "Seiryu", with open space perfect for relaxing

9. The open-air spa where you can enjoy the natural scenery

7. 男用大浴场"溪流"，设计稳重、宽敞，温泉量丰富。浴场内设有按摩浴缸、泡泡浴缸等，于清澈之水洗涤累积的疲劳

8. 女用大浴场"清流"，空间明亮、开放，可充分舒缓身心疲劳

9. 露天温泉，解除身心束缚，享受自然美景

10. Fish pond

11. The teahouse

12. The spacious and calm guestroom in pure Japanese style, making guests feel at ease resting here

10. 大厅旁的锦鲤鱼池塘

11. 茶室

12. 宽敞稳重的日式空间，让客人自在休憩

Hotel Hana-Isawa
花石和温泉饭店

Location: Matsumoto 1409, Isawa-cho, Fuefuki City, Yamanashi Prefecture, 406-0021, Japan

Spa Name: Isawa Onsen

Area: 9,550 m²

Quantity of Guest Room: 69

Quantity of Spa Bath: 5

Other Facilities: private open-air spa, chartered spa

地址：日本山梨县笛吹市石和町松本1409

温泉名：石和温泉

面积：9,550平方米

客房数：69

温泉浴池数：5

其他设施：专属露天浴池、包租浴池

It only takes five minutes to the Hotel Hana-Isawa, located in Matsumoto, Isawa-cho, Fuefuki City, Yamanashi Prefecture by JR (Japanese Railway). With the concept of "enjoy your easy and cosy spa time", Hotel Hana-Isawa features overflowing onsen with a hot spring source coming from its properties, the natural flow with abundant hot water. There are two types of rotemburos: "Retemburo Hoshikuzu No Yu" and "Retemburo Gessho No Yu", with the different themes of gentle and strong style. The most popular onsen that continuously overflows from the bath tub is a proprietary onsen known for its soft and smooth properties. It has a refreshing effect to skin. The glass walls of the large public bath presents with a feeling of liberation and openness. Various choices of guest rooms are offered, such as Japanese-style rooms, Western and Japanese-style rooms and "Karei" Bungalow with Rotemburo, accompanying with guests to enjoy the luxury leisure time.

In Japan, the introduction of onsen is always coupled with that of the Japanese cuisine. With the local elements of Yamanashi Prefecture, Kaiseki Ryori is created by experienced chef. The traditional Japanese cuisine is brought in courses one by one, into which the chef has incorporated the careful choices of ingredients. The semi-private dinning room "Hanakomichi" has a high sense of privacy and provides a moment of elegance. The exquisite

cuisine is made of seasonal ingredients. The cuisine contents and even the crockery are changed with seasons.

花石和温泉饭店位于山梨县笛吹市石和町松本温泉乡，乘日本铁路JR车到石和温泉后，只需步行5分钟。该温泉饭店以"悠闲与滋润的时光（泡温泉）"为设计主题思想。花石和温泉饭店的最大特点，就是引自馆内源泉的"溢出温泉"（也叫"自喷温泉"）拥有丰富的水量。被称为"星宵温泉"、"月宵温泉"的露天浴池分别体现柔和与力量。不断涌出的温泉水水质温润爽滑，是最受好评的自涌式温泉，加之大浴场采用玻璃窗设计，具有良好开放感。这里有日西式结合客房和设有露天温泉的独栋客房"花龄"等，以多种方式让客人度过悠闲的奢华时间。

介绍日本的温泉离不开日式料理，花石和温泉饭店厨师以精湛厨艺制作出富有山梨县地方特色的会席料理。由主厨亲选食材的会席料理，以舒缓的节奏分批上菜。半包厢形式的餐厅"小上座敷花小路"提供高雅而不受干扰的用餐环境。料理使用时令食材，菜式和餐具也会随季节而变动。

1. The entrance

2. The façade of the hotel, which re-opened after the renovation in 2002

3. The open-air spa "Rotemburo Gessho No Yu"

4. Spa "Afureide no Yu" with the onsen that continuously overflows from the bath tub, a proprietary onsen well-known for its very soft properties

1. 入口

2. 外观，2002年全馆整修后重新开业

3. 露天月宵温泉

4. "溢出温泉"不断从浴缸涌出的温泉水水质柔软，是深受好评的自涌式温泉

5. Guest room "Karei" Bungalow with Rotemburo

5. 带有露天温泉的独栋客房"花龄"

1. 入口

2. 外观，2002年全馆整修后重新开业

3. 露天月宵温泉

4. "溢出温泉"不断从浴缸涌出的温泉水水质柔软，是深受好评的自涌式温泉

6. The open-air spa "Rotemburo Hoshikuzu no Yu"

7. The private dinning room "Hana-Isawa", the restaurant with private rooms (for up to 4 persons) equipped with "hori-kotatsu" (low heated table with legroom built into the floor). Guests can spend a blissful moment savouring each of the dishes one by one

8. The dinning room "Hanakomichi" is semi-private dinning room, equipped with "hori-kotatsu", which has a high sense of privacy and provides a moment of elegance

6. 露天星宵温泉

7. 包间餐厅"花料亭"。全包厢的花料亭（最多4人）采用炕式暖桌坐位，一道道怀石料理，让游客回味无穷

8. 餐厅"花小路"（Hanakomichi）为半开放包间设计，配有坑式餐桌。半包厢形式的餐厅"小上座敷花小路"提供高雅而不受干扰的用餐环境

7

8

9. Japanese-style rooms, a warm and cosy room with an area of 16.5 square metres

10. "Karei" Bungalow with Rotemburo designed with a height of a patio, offering a feeling of leisure and openness

11. The private dinning room equipped with "hori-kotatsu"

9. 日式客房，16.5平方米宽敞、高雅的日式客房，温馨舒适

10. 独栋客房"花龄"，天井设计营造开放悠闲感

11. 包间餐厅，配有坑式餐桌

Kasugai View Hotel

春日居VIEW饭店

Location: Shizume 178, Kasugai-cho, Fuefuki-City, Yamanashi Prefecture, 406-0021, Japan

Spa Name: Isawa Onsen

Area: 10,367 m²

Quantity of Guest Room: 71

Quantity of Spa Bath: 4

Other Facilities: private open-air spa, rentable spa

地址：日本山梨县笛吹市春日居町镇目178

温泉名：石和温泉

面积：10367平方米

客房数：71

温泉浴池数：4

其他设施：专属露天浴池、包租浴池

Nestled in Isawa Onsen in central Kofu Basin, Yamanashi Prefecture, Kasugai View Hotel features the statue of kind and friendly Ikkyu san, relaxing enviroment and delicious food. At the entrance of the hotel, the statue of the kind and friendly monk, Ikkyu san is welcoming guests from the world. Walking inside, the reception hall decorated with large pieces of glass with a height to the patio is full of gentle and warm sunlight. There are three Japanese and Western-style guest rooms among the 71 guest rooms with spacious and free interior space. Covering an area of 8 tatami mats to 12.5 tatami mats (13 square metres to 23 square metres), the Japanese guest rooms with typical Japanese elements and interest are all paved with Japanese tatami. The open-air spas boast alkaline simple hot spring with plentiful volume. You can enjoy bathing while admiring the pastoral scenery with rustic interest and grand openness.

The Fuefukigawa area in Kofu Basin, the location of Kasugai View Hotel, has long been crowned as "Fruit and Onsen Town", where grows a large area of peach trees and vineyards. It enjoys the largest output of peaches and grapes in Japan. In summer, tourists can take part in the self-help peach picking events, whilst grape picking festival in autumn.

Buffet dinner here provides crabs, freshly-baked beef, kebabs, etc. The cosy and tranquil Japanese dinning room provides more than 40 kinds of Japanese and Western food as well as the famous refined wine of Yamanashi Prefecture.

春日居VIEW温泉饭店位于山梨县甲府盆地中央的石和温泉乡，设计上着重表现"温厚亲切的一休、放飞心情、尽情品尝"。饭店入口处迎接客人的是温厚亲切的一休。迎客大厅由直达天井的大面积玻璃装饰，室内洒满柔和温暖的阳光。全部71间客房中，有3间为日式西式混搭风格，室内空间宽敞舒适，充满和式意趣的和室铺设8~12.5帖（13~23平方米）日式榻榻米。春日居VIEW温泉饭店露天温泉水量丰沛，其水质为碱性单纯泉质，躺在露天温泉浴池里欣赏田园景色野趣满满、自在畅快。

春日居VIEW温泉饭店所在的甲府盆地的笛吹川流域是著名的"水果和温泉之乡"，这里有成片的桃林和葡萄园，桃和葡萄的产量全日本第一。夏天，游客可以参加这里的桃园自助式采桃活动；秋天，则成为了摘葡萄的节日。

高级自助晚餐提供不限量自由享用的螃蟹，同时还提供现烤牛肉、油炸肉串等。和式风格的餐厅温馨安静，包括西餐在内约40余种食品，此外，还可享用山梨县的名产——特制葡萄酒。

1. The façade

2. The reception hall decorated with large pieces of glass with a height to the patio in the hug of the gentle and warm sunlight

3. The hall

4. The bathing pool

1. 外观

2. 由直达天井的大面积玻璃所装饰的迎客大厅洒满柔和温暖的阳光，宽敞、开阔

3. 大厅

4. 浴池

5. The open-air spa, enjoy bathing while feeling the breezes of the seasons in the rotemburo, the pride of the hotel

5. 酒店引以为傲的露天温泉，可在季节微风吹拂之下享受泡温泉的乐趣

6. The banquet hall with an area of 48 tatami mats (78 square metres), which can accommodate 48 persons

7. The dinning room in which both breakfast and dinner are buffet-style

6. 宴会厅铺设榻榻米48帖（78平方米），可容纳48人

7. 餐厅，提供自助早、晚餐

8. The Japanese-style guest room, which is recommended to those guests who wish to relax

9. The Western and Japanese-style guest room with a choice of 10-square-metre and 16.5-square-metre

10. The Western-style double-bed, economical choice for business or couple trip

8. 日式客房，适合想放松心情好好休息的游客

9. 日西式客房，分为榻榻米空间10平方米及16.5平方米的两种类型

10. 西式双人客房。不论商务或情侣、夫妻履行，都是经济实惠的好选择

Southern Cross Resort Hotel
南极星度假村

Location: 1006 Yoshida, Ito City, Shizuoka Prefecture, 414-0051, Japan

Spa Name: Ito Onsen

Area: 720,000 m²

Quantity of Guest Room: 55

Quantity of Spa Bath: 2

Other Facilities: private spa bath

地址：日本静冈县伊东市吉田1006

温泉名：伊东温泉

面积：72万平方米

客房数：55

温泉浴池数：2

其他设施：专属温泉浴池

Seated in Yoshida, Ito City, Shizuoka Prefecture, Southern Cross Resort Hotel is surrounded by charming sceneries and beautiful landscapes such as Mount Fuji on its north, the seven Izu Islands on the south, Sagami Bay on the east and Mount Amagi on the west. Covering a total area of 720,000 square metres, it is equipped with an 18-hole championship golf course and 6-hole golf course.

The onsen of spa in the Southern Cross Resort Hotel comes from a source within its own properties and there are also water floating activities designed in the swimming pool with hot spring. In spring, blooming flowers can be found everywhere in the hotel. Different kinds of flowers changed with seasons, decorating each and every corner of the golf course. Wandering in the golf course with the aroma of natural flowers and grass makes you a part of the whole nature.

Izu Peninsula is a scenic spot with the blessed mountains and sea. All ingredients are locally produced, including the fresh marine fish caught in the coastal waters. Guests are also provided with Guri tea from Izu and Wasabi from Amagi.

南极星温泉度假村位于静冈县伊东市吉田，北临灵峰富士，南边为伊豆七岛，东为相模湾，西朝天城连山，四面景色明媚、风光秀丽。度假村总占地面积为72万平方米，内设18洞国际标准高尔夫球场和6洞短球场。

南极星温泉度假村内的温泉大浴场所使用的温泉来自度假村内涌出的源泉。在度假村的温水游泳池内可以进行放松身心的水中漂浮运动。 到了春天，度假村内处处樱花盛开。随着季节变换，标准高尔夫场内到处开放着不同种类的鲜花，漫步在场内的步行道上，呼吸花草气息，使人整个身心都融入大自然。

度假村所在的伊豆半岛是一个拥有山和海所恩赐的富饶资源的观光胜地。度假村所使用的食材全部来自当地，包括在伊豆半岛近海所捕获的新鲜海鱼。度假村还为客人准备了伊豆的铭茶和天城的山葵（辣根）。

1. The façade

2. The façade of Southern Cross Resort Hotel

3. Night view of open-air spa

4. The swimming pool with warm water is available all year round

1. 外观

2. 外观，南极星温泉度假村

3. 露天温泉夜景

4. 全年可使用的温泉游泳池

5. The view of Mount Fuji in the golf course

5. 从高尔夫球场可以眺望到富士山

6. The club with the amenities such as the bar and the karaoke hall

7. The dinning room serving Italian set meal and Japanese Kaisseki Ryori for dinner

8. The guest room mixed Japanese and Western style

6. 俱乐部，设有酒吧和卡拉OK厅

7. 餐厅，晚餐提供意大利套餐和日式会席料理两种

8. 日式与西式风格相融合的客房

8

9. The guest room for business use

10. The standard guest room

11. The Western-style guest room

9. 商务客房

10. 标准客房

11. 西式风格客房

Hotel Laforet Nasu
那须 Laforet温泉旅馆

Location: 206-959, Yumoto, Nasu-machi, Nasu-gun, Tochigi Prefecture, Japan

Spa Name: New Nasu Spa

Area: 28,027 m²

Quantity of Guest Room: 118

Quantity of Spa Bath: 4

地址：日本栃木县那须郡那须町汤本206－959

温泉名：新那须温泉

面积：28027平方米

客房数：118

温泉浴池数：4

Shiobara Onsen Town on the upstream of Houkigawa is just on the southwest of Nasu Highland. It is crownded as "Shiobara 11 Onsens" for the 11 onsens there. In autumn with red maple leaves, the surrounding area of the stream presents beautiful and colourful views such as the Momijidani suspension bridge with a total length of 320 metres at the entrance of the onsen town, the Kaiko suspension bridge along the sightseeing pedestrian street and the Kaiko waterfall, which are the ideal places commanding a view of the stunning scenery.

Hotel Laforet Nasu is located on Nasu Plateau, north of Tochigi Prefecture on the upstream of Naka River. The plateau area rolling up and down at the foot of Nasu Volcano is dotted with onsens such as Hakoneyumoto Onsen, Daimaru Onsen, which make up Nasu Onsen Town. Hotel Laforet Nasu is located on Nasu Plateau with stunning natural landscape in all seasons. Tourists can catch sight of beautiful scenery of continuous mountains of the grand Nasu Plateau, namely, the red maple leaves in autumn and the emerald green branches in spring, whether from the guestrooms or the bath area. The soft and smooth hot spring of the spa, in milky white, has the special efficacy for skincare. The hotel is highly proud of the Western-style cuisine cooked with natural ingredients and the traditional banquet cuisine.

The interior space is equipped with facilities such as conference room, banquet hall and arena. There are also a large number of surrounding recreational facilities such as the tennis court, golf court and ski resort and attractions of various kinds near Nasu Plateau such as Nasu Gallery, Nasu Zoo as well as Nikko Toshogu, the world heritage.

那须偏西南的盐原温泉乡位于帚川上游溪谷旁，因有11处温泉，故有"盐原11温泉"的昵称。秋天红叶季节的溪谷景色优美。温泉乡入口处的枫叶谷大吊桥全长320米，徒步游览时吊桥、瀑布等都是眺望美景的好地方。

那须 Laforet温泉旅馆位于那须高原、栃木县的北部那珂川的上游。那须火山脚下绵延起伏的高原地带上分布着汤本温泉、高熊温泉、大丸温泉等等，形成了那须温泉乡。从旅馆客房或者浴场望出去，就能欣赏到那须高原雄伟的连绵群山，秋天红枫叶满山，春天满眼翠绿新枝。馆内温泉泉质柔顺润滑，呈天然乳白色，对美化肌肤有特别功效。利用天然食材烹制的西式料理和传统的宴会料理是这里引以为傲的特色之一。此外还设有会议室、宴会厅、竞技场等公共设施，周边娱乐设施场所也很多，有网球场、高尔夫球场、滑雪场等。附近其他那须高原观光景点有那须美术馆、那须动物园，还有世界遗产日光东照宫等。

1. The façade of the hotel located on Nasu Plateau with the stunning natural landscape in all seasons

2. The night scene of the façade

3. The indoor large public bath with hot spring in milky white, where you can overlook Nasu Plateau out of the window

4. The conference room

1. 酒店外观，酒店地处四季景观秀美的那须高原

2. 酒店外观夜景

3. 室内大浴池，温泉水呈天然乳白色，从窗口望出去可俯视那须高原

4. 会议室

5. The spacious hall

5. 宽敞的大厅

3

4

6

6. The dinning hall

7. The Japanese-style guest room with an area of 42.7 square metres

8. The interior space of the independent house with an area of 75.5 square metres, in which there is a spacious living room for group events

6. 餐厅

7. 日式风格的客房，面积为42.7平方米

8. 独栋小屋室内，总面积为75.5平方米，有宽敞的起居室，适合团体活动

9. The guest room, mixing Japanese and Western style

10. The Western-style guest room in the main colour of brownish gold

11. The luxury guest room

9. 日式、西式风格相融合的客房

10. 以金茶色为基调的西洋风格客房

11. 豪华客房

Laforet Club Hotel Naka-Karuizawa
轻井泽 Laforet俱乐部酒店

Location: 4339 Nagakura, Karuizawa-machi, Kitasaku-gun, Nagano, 389-0111, Japan

Spa Name: Karuizawa Shiotsubo Onsen

Quantity of Guest Room: 84

Quantity of Spa Bath: 4

地址：日本长野县北佐久郡轻井泽町长仓4339

温泉名：轻井泽盐壶温泉

客房数：84

温泉浴池数：4

It takes only one and a half hour by trolley bus from Tokyo to the famous Karuizawa Resort located in Kanto, Japan. The hotel buildings are surrounded by green trees, which make you feel like staying in a fictitious land of peace and happiness. It features double-bed rooms and an annexe which can accommodate 10 persons. The simple quality of the hot spring is particularly effective in curing neuralgia, digestive system disease and alleviating fatigue.

The cuisine prepared for guests includes authentic French Cuisine, roast meat, etc. There are also a large number of tourist attractions nearby such as old Karuizawa Ginza, Karuizawa Shopping Centre, Shiozawa Lake and the stunning scenery of red maples in Kumo ba Pond.

It is a hotel amidst the surrounding trees in the full-dodied Karuizawa style. Various facilities such as onsens, beauty salons and golf courses can be found here. Tourists can spend unforgettable leisure time here with the extreme elegant facilities from the tranquil and cosy guestrooms to the dinning room and the spa.The classic Karuizawa cuisine matches perfectly with the fresh and cool Karuizawa dinner. The local ingredients in fresh colours are cooked elaborately by the cooks. Tourists may feel satisfied with whether the entrée or the jardinière.

酒店位于日本关东著名的避暑胜地轻井泽，距离东京1.5小时的电车路程。酒店建筑群四周绿树成荫，宛如世外桃源。本酒店的客房以双人套房为主，同时设有可以同时容纳10人的别庄。温泉为单纯泉质，对神经痛、消化系统疾病、疲劳等有特殊功效。

酒店为客人提供的料理有正宗法国料理、铁板烤肉等。周边的旅游景点众多，如轻井泽小道、轻井泽购物中心、盐泽湖风光、云场池红枫叶奇景等。

这座建于林中、具有浓郁轻井泽风格的酒店温泉，美容、高尔夫等设施齐全，游客可以在这里度过无比轻松的假日。从舒适静谧的客房到餐厅，温泉优雅至极的设施让游客在轻井泽度过最难忘的休闲时光。本地产的时鲜蔬菜经过厨师们的精心烹饪料理，从主菜到配菜都让人无法挑剔，是经典的轻井泽料理。

1. The façade of the hotel which enjoys full openness in the beautiful environment of blue sky and green grass

2. The façade of the independent annexe with adequate privacy in the embrace of the nature

3. The reception lobby made from marble, making guests feel comfortable

4. The vestibule

1. 酒店外观，绿草蓝天，开放感十足

2. 独立别庄外观，自然环抱，隐秘宁静

3. 酒店大理石前台大厅

4. 酒店玄关

5

5. The streamlet in the garden of the hotel

5. 酒店园区内的小溪流

3

4

6

6. The façade of the dinning hall   6. 餐厅外观

7. One corner of the open-air spa   7. 露天温泉一角

8. The large spa bath with a river running outside the window just like a white ribbon

9. The luxury double guest room

10. The standard double guest room

8. 温泉大浴场，窗外的水流如同白丝带般缓缓流淌

9. 豪华双人客房

10. 明亮温暖的标准双人客房

Laforet Club Hotel Ito

伊东 Laforet俱乐部酒店

Location: 2-3-1 Shishido, Ito City, Shizuoka Prefecture, 414 – 0004, Japan

Spa Name: Ito Onsen

Area: 3,800.47 m²

Capacity: 360

Quantity of Guest Room: 84

Quantity of Spa Bath: 7

Other Facilities: rentable spa

地址：日本静冈县伊东市猪户2丁目3番1号

温泉名：伊东温泉

面积：3800.47平方米

容纳人数：360

客房数：84

温泉浴池数：7

其他设施：包租温泉

It is only eight minutes' walk from Ito Station to Laforet Club Hotel Yitou, which locates in the east of Izu Peninsula, only 100 kilometres away from Tokyo. Located next to the Ito Orange Beach of Ito, it enjoys a tranquil environment and is very suitable for relaxing.

Simple but comfortable rooms are available with wireless network, which is convenient for searching information about the sightseeing facilities and scenic spots. There are so many tourist attractions nearby such as Jogasaki Beach, Izu Kogen, Sakura Town and Omuroyama.

The cuisine served is mainly seafood dishes, among which the marine fish are from the nearby Ito Port. The hot spas use the colourless and transparent simple alkaline spring water which has unique efficacy in curing neuralgia, muscle pain and alleviating fatigue, suitable for those tired guests after long-term travel.

酒店位于距离东京100公里左右的伊豆半岛东部，从伊东站步行只需要8分钟。酒店邻近伊东的黄色海滩，但周围环境非常安静，很适合疗养身心。

酒店各个客房简洁舒适，设有无线网络接入，在客房或者酒店内各处都可以实现无线上网，方便检索酒店附近的观光设施和景点信息。酒店附近有名的旅游景点很多，如城崎海岸、伊豆高原、樱花之乡、大室山等。

酒店主要提供海鲜料理，所使用的海鱼是直接从附近的伊东港进货的。温泉的泉质为单纯温泉，成弱碱性，无色透明，对神经疼痛、疲劳、肌肉疼痛等症状有特别功效，最适合长期旅行身体疲惫的客人。

1. The façade
2. The fountain in the atrium
3. The rentable spa for family

1. 酒店外观
2. 中庭喷水池
3. 家庭包租浴池

4. The wooden spa pool
4. 木制温泉浴池

5

5. The rockstone-made spa pool, named "Kenjin no Yu", has a long history

6. The Ball Jointed Doll show on "Hinamaturi" (The Daughter's Festival) on March 3rd

5. 岩石砌造浴池，名为"健身汤"，具有悠久历史

6. 3月3日日本女儿节"雏祭"（Hinamaturi）人偶展示

7

7. The hall   7. 大厅

8. The dinning hall   8. 餐厅

9. The guest room, mixing Japanese   9. 融合日式、西洋风格的客房
and Western style

Zagyosoh
坐渔庄温泉旅馆

Location: 1741, Yawatano, Ito City, Shizuoka Prefecture, Japan

Spa Name: Ukiyama Hotspring Village

Area: 13,200 m²

Quantity of Guest Room: 20

Quantity of Spa Bath: 4

Other Facilities: private open-air spa, rentable spa

地址：日本静冈县伊东市八幡野1741

温泉名：伊东浮山温泉乡

面积：13200平方米

客房数：20

温泉浴池数：4

其他设施：专用露天浴池、包租温泉

Located in Ukiyama Hotspring Village, Ito City, Zagyosoh is a Japanese-style hotel full of the lasting appeal and aroma of the harbour city. It takes only two hours from Tokyo by trolley bus. Founded in 1968 with a covering area of 13,000 square metres, it boasts 20 elaborately decorated guest rooms. You can catch sight of the hundred-year-old peach trees, like the dotted stars in the sky, on the beautiful Kinosaki Coast, the waterfall outpouring from the mountains inside the natural garden formed by interweaved lava in thousands of postures, and the six spa facilities in different styles such as Teien-Roten Buro, Fuhketsu Roten Buro and Tenjyou Ocean Front Open-air Spa. The simple alkaline hot spring is so gentle and smooth that bring joyness of spa.

All guest rooms are in the typical Japanese style with the fragrance of the natural cypress, the Japanese aroma of tatami and the comfortable feeling of the natural gardens, which make guests fully intoxicated.

Japanese Kaiseki is served for dinner with the main ingredient of marine fish. Guests can not only taste the natural flavour by enjoying the elaborately cooked cuisine, but also feel the traditional Japanese hospitality.

坐渔庄温泉旅馆坐落在伊东市的浮山温泉乡，这是一座充满港口城市风情的日式酒店，距东京2小时电车路程。坐渔庄温泉旅馆始创于1968年，占地约1.3万平方米，共有20间精心装修的日式房间。从客房望去可见风光秀丽的城之崎海滨沿岸，百年山桃树像繁星密布；在千姿百态的、熔岩交织而成的自然庭园中还能看到从山间倾泻的瀑布。这里有观海天上露天温泉浴池，风穴露天温泉浴池等6个不同风格的温泉浴池，旅馆内的温泉为弱碱性，泉质柔顺润滑，让游人可以享受到泡温泉的乐趣。

旅馆的建筑设计采用纯和式风格。天然桧木散发出淡淡的清香，榻榻米代表和式风情与自然庭园的舒适，让游客们无比沉醉。

为客人准备的晚餐为"创作和风怀石料理"，以新鲜海鱼为主，精心制作而成的美味料理不仅让游客享受到大自然的美味，同时还体会到旅馆传统的"款待之心"。

1. The vestibule of Syoukintei, the independent guest room of the annexe

2. The vestibule

3. The sunset view

4. The open-air spa "Fuketsu"

1. 别邸（独立客房）"松琴亭"的玄关

2. 玄关

3. 夕阳美景

4. 风穴露天浴池

5. The garden of the eastern building inside the typical Japanese-style hotel with natural gardens and interesting spa facilities on the 13200-square-metre site

5. 东馆庭园，在13200平方米的区域内设有自然庭园、充满趣味的温泉设施

6. The interior garden of the guest room "Hatsune"

7. The open-air spa "Teien", the stone open-air bath in the green garden which makes your bath so close to nature

8. The chartered open-air spa commanding a panoramic view of the beautiful landscape of Sagami Bay and Izu Oshima

6. 客房"初音"的室内庭院

7. 庭园内岩石露天浴池，于绿色环绕之中

8. 包租露天温泉（观海天上露天浴池），可以充分领略到相模湾的美景与伊豆大岛的风貌

9. The verandah of Syoukintei, the independent guest room of the annexe

10. The indoor spa "Ukibune", high-grade cypress made, in which you can admire the peach trees

11. Terrace stretching out to the garden area

9. 别邸（独立客房）"松琴亭"的长廊

10. 室内大浴池"浮舟"，高级桧木打造，从浴池可以观赏到室外的百年山桃树木

11. 露台直接延伸至庭园

14

12. The large and spacious banquet hall equipped with a stage

13. The guest room of Syoukintei, the independent guest room of the annexe with the exclusive open-air spa

14. The eastern venue, a VIP room in typical Japanese style with a private garden and an open-air spa

12. 宽敞大宴会厅内配有舞台

13. 别邸（独立客房）"松琴亭"，带有专用的露天温泉

14. 东馆，带有专属庭园和露天温泉的贵宾客房，纯日式风格

Yokikan

阳气馆

Location: 2 – 24, Suehiro-cho, Ito City, Shizuoka, 414-0015, Japan

Spa Name: Ito Onsen

Area: 3,000 m²

Quantity of Guest Room: 19

Quantity of Spa Bath: 3

Other Facilities: private open-air bath

地址：日本静冈县伊东市末广町2 – 24

温泉名：伊东温泉

面积：3000平方米

客房数：19

温泉浴池数：3

其他设施：专属露天浴池

Located in Suehiro-cho, Ito City, Yokikan had spent the 100ᵗʰ anniversary by July, 2010, since its foundation in Meiji 43 (1910 AD). The one-hundred-year-old hotel takes most pride in its spa facilities (one open-air spa and two indoor spas). It was designed in the elegant environment with beautiful landscapes, occupying the favourable natural terrain of the high hills behind. The mountain trains can take you to the well-known open-air spa and the annexe. The hotel takes pride in its open-air spa with hot springs gushed directly from its properties. Guests may feel like being inside a mountain when entering the central lobby where the waterfall falls vertically among the luxuriant and green trees, creating a dynamic and declicate picture. The two natural hot springs with their own water qualities have different effects. Guests can overlook the Ito Sea and the townscapes of Ito City. It is so quiet and peaceful during nighttime when stars are dotted the whole sky.

Since its foundation, Yokikan has retained its tea house architectural style. There are all together 19 guestrooms built in traditional Japanese style with their own characteristics in three pure Japanese-style buildings on the hillside including the main building, the annexe and Suehirotei. The guests can recuperate themselves in the natural world, being a part of it.

Ito cuisine is always famous for its sea food. Various kinds of cuisine and carefully-made delicacies are cooked with the seasonal local ingredients and seafood selected from the offshore.

阳气馆位于静冈县伊东市末广町，始建于明治四十三年（1910年），2010年7月迎来100周年。创业百年的日式阳气馆最引以为傲的是温泉浴池（露天温泉1处，室内浴池2处）。它巧妙地利用后山高岗的自然地形条件设计而成，环境幽雅、风景秀丽。乘坐旅馆内的登山电车可以很方便地将游人送到著名的野地温泉浴池及别馆。阳气馆中厅里，自有的源泉垂流式露天大浴池中温泉瀑布垂直落下，灵动俊秀；周围树木茂密，郁郁葱葱，恍如置身深山。阳气馆自有源泉两处，泉质不同，有不同的功效。从露天温泉可以俯瞰伊东大海和伊东市井风貌。夜幕降临时星空缭绕，一派安静祥和的气氛。

阳气馆自创业以来一直保留着古老的茶室建筑风格。坐落于大山腹地的3栋建筑是纯日本式建筑，别馆、本馆、末广亭加起来共有19间客房。客房采用传统设计风格，品位各异。游客在自然中休养，与周围的风景融为一体。

伊豆料理以海鲜而闻名。旅馆精选近海海鲜和当地特有的"时令食材"，为客人们提供充满季节感的各色料理。

静山荘
政府登録
国際観光旅館
陽気館

1. The hall – the hotel has retained the tea house architectural style for 100 years since its foundation

2. The vestibule, recuperating in the natural world

3. The dynamic and delicate waterfall falls down in the central lobby

4. The open-air spa commanding a panoramic view of the sea and the whole city

5. The night scene of the spa

1. 大厅。创业100周年以来一直保留着古老的茶室建筑风格

2. 玄关，能够感受到身心在自然中得到休养

3. 中厅，温泉瀑布垂直落下，灵动俊秀

4. 可以一览海洋和全市的露天浴池

5. 温泉夜景

6. The vestibule of the pure Japanese-style hotel

6. 纯日式旅馆玄关

7

7. The outdoor swimming pool which is only open in summer

8. The after-bath rest area

9. The corridor of "Karei" Bungalou

7. 户外游泳池，限定夏天开放

8. 浴后休憩处

9. "离"（别邸独立客房）的走廊

10. "Karei" guest room

11. The standard guest room

12. The guest room inside the main building

10. "离"（别邸独立客房）

11. 标准客房

12. 主建筑内的客房

Inatori Gin suiso

稻取温泉银水庄

Location: 1624-1, Inatori, Higashiizu-cho, Kamo-gun, Shizuoka Prefecture, 413-0411, Japan

Spa Name: Inatori Onsen

Capacity: 636 persons

Area: 15,370 m²

Quantity of Guest Room: 117

Quantity of Spa Bath: 4

Other Facilities: outdoor spa with indoor service

地址：日本静冈县贺茂郡东伊豆町稻取1624－1

温泉名：稻取温泉

容纳人数：636

面积：15,370平方米

客房数：117

温泉浴池数：4

其他设施：带室内服务的露天温泉

Situated in Inatori, Higashiizu-cho, Kamo-gun, Shizuoka-ken, Inatori Ginsui boasts the beautiful scenery of the southland for its ideal location on the east coast of Izu Peninsula close to the Pacific Ocean. Since its foundation in 1957, it has been cherished so much by Japanese people that it has ranked 1st for 16 years in the activity of "Top 100 Japanese Onsens Elected by Experts" host by Japan Travel News Agency. Focusign on the Japanese tradition of hospitality, Inatori Ginsui is a modern Japanese-style hotel welcoming tourists with its own elegance and politeness, showcasing the quintessence of Japanese service.

All guestrooms are oceanfront ones where guests can overlook Izu Shichito Islands and the long beautiful coastline of Izu Peninsula.

There is another outstanding design in Inatori Ginsui – all baths including "Ayame Spa" for men, "Kiou Spa" for women and various open-air bath pools that were designed as "Miyabi kouji", Japanese Ryotei and tea house, and even every rafter were made of Kitayama cedar, which look so luxurious and elegant.

The food served here are the elaortately cooked Iso Ryori (Seafood Feast), coupled with the delicate appearance and considerate service. There are Chinese-speaking waiters and menus in Chinese.

稻取温泉银水庄位于静冈县贺茂郡东伊豆町稻取，地处伊豆半岛东海岸，濒临太平洋，充满南国绮丽的风光。稻取温泉银水庄创建于1957年，至今深受日本人的喜爱，在日本旅游新闻社主办的"专家评选百家日本一流温泉旅馆"活动中，连续16年排名第一。银水庄非常注重日本传统的服务精神——让客人感受到"热情招待之心"。它是一家以优雅姿态迎接游人的现代化日式旅馆，向游客充分展示了日式服务的精髓。

稻取温泉银水庄所有客房均为海景房，可以远眺伊豆七岛和伊豆半岛漫长美丽的海岸线。

稻取温泉银水庄另一个特别之处是菖蒲（男）、姬樱（女）两大浴池，此外，还有洋溢着京都风情的"雅小路"、日式料亭和茶室，甚至每一根椽子都使用了北山衫，奢华考究。

稻取温泉银水庄日式料理是精心烹制而成"矶会席料理"（海鲜宴席），款式精致，烹调手法多样，并且服务细致周到，有会说中文的服务人员，并专门提供了中文菜单说明。

2

3

1. The façade

2. The Japanese Ryotei

3. The night scene of the open-air spa with sea-views

4. The guest room "Zuiun", equipped with high-grade cypress spa

5. The Japanese and Western-style guest room with open-air bath

1. 酒店外观

2. 料亭

3. 夜景

4. 客房"瑞云"，配有专用高级桧木温泉

5. 日式与西式风格融合的客房，带专用露天浴池

6. The regular guest room of 11 to 14.5 tatami mats (18 square metres to 24 square metres), with a wide corridor

6. 普通客房，面积榻榻米11~14.5帖（18~24平方米），附加宽走廊（广缘）

4

5

Dougashima New Ginsui
堂岛新温泉银水庄

Location: 2977–1, Nishina, Nishi-Izu-Cho, Kamo-gun, Shizuoka Prefecture, 410-3514, Japan

Spa Name: Inatori Onsen

Area: 49,950 m²

Quantity of Guest Room: 123

Quantity of Spa Bath: 4

Other Facilities: open-air spa with indoor service

地址：日本静冈县贺茂郡西伊豆町仁科2977–1

温泉名：稻取温泉

面积：49,950平方米

客房数：123

温泉浴池数：4

其他设施：带室内设施的露天温泉

As a sister hotel of Inatori Ginsui, Dougashima new Ginsui is a large-size spa hotel in Dojima, Nishi-Izu-Cho, Kamo-gun, Shizuoka-ken, the famous scenic spot of Nishi-Izu Peninsula. With the basic concept of modern Japansese leisure style, it is a new modern hotel with all-round facilities and top amenities. All guest rooms are ocean fronted, which have a commanding view of the stunning scenery of Dojima and the everchanging coast with ebbs and flows. There are open-air spas, sauna baths and rest areas, where the guest can overlook the greatest scenery of Suruga Bay and Sanshirojima Island. The hot spring swimming pool and other kinds of playgrouns can also be found on the ground floor.

From early October to the first ten days of March next year, guests can admire the stunning scenery of the harmonious amalgamation of the sea and the sky, while enjoying wine drinking (free of charge) on the terrace of Dougashima new ginsui. There are also a large number of tourist attractions and recreational facilities nearby, taking the Dojima Ship as an example, by which the guest can admire the natural picturesque sea cave. Numerous scenic spots and recreational facilities can be seen along the Izu Coastline to Mount Fuji.

The cuisine here is carefully made by the professional Japanese chefs. The menus feature Iso Ryori with abalone or Ise Lobster. There are also Chinese staffs who can offer services in Chinese and moreover, menus in Chinese are available.

堂岛新温泉银水庄与稻取温泉银水庄是姊妹馆，位于静冈县贺茂郡西伊豆町，是西伊豆半岛风景名胜地"堂岛"的一家大型温泉酒店。它的设计体现的是一种轻松和风主题，也是一家融合了南欧风情的、现代化新式温泉度假大酒店，设施齐备，各方面都勘称一流。客人在每间客房都可以眺望到堂岛的美丽风光，观察潮长潮落变幻无穷的海岸。这里设有露天浴池、桑拿浴池和宽敞的休息区，泡温泉的同时可以遥望骏河湾和三四郎岛美景。酒店一层设有温泉游泳池和多个游乐场。

每年10月上旬开始，一直到来年3月上旬，在堂岛新温泉银水庄旅馆的露台上，可以边饮用免费提供的葡萄酒，边欣赏夕阳西下、海天一色的绝色美景。旅馆周边有很多著名的旅游景点和游玩方式：乘坐"堂岛游览船"可以观赏天然的、奇观如画的海上洞窟。从西伊豆海岸线开始，一直到富士山，有数目众多的游玩观光场所。

堂岛新温泉银水庄餐饮由专业日本料理师精心制作，以矶料理为主。客人根据所定菜单的不同，可以享受到鲍鱼或者伊势大龙虾。本馆有中国职员，可以提供中国语服务，并且备有中文专用料理菜单。

1. The façade of the hotel

2. Dojima, also called "Matsushima in Izu" – splendid landscape combining the island on the sea and the coast with cliffs

3. The lighting at the vestibule

4. The exterior scenery

1. 外观

2. 堂岛又称"伊豆之松岛"，碧海与悬崖峭壁的海岸形成一道壮丽景观

3. 玄关夜间照明

4. 外景

5. The façade

5. 外观

6. The hall   6. 大厅

7. The view of the hall from the above   7. 大厅俯瞰

8. The reception hall   8. 接待大厅

7

8

9. The open-air spa

10. The indoor observatory large bath

11. The hall commanding a view of Nishi-Izu

9. 露天温泉

10. 宽阔的室内观景浴池

11. 大厅尽享西伊豆夕阳景色

12

12. The standard Japanese-style guest room with a room area of 14.5 tatami mats and a corridor

13. The Japanese-style guest room

14. The Western-style guest room

12. 标准和式客房，14.5帖＋宽走廊（广缘）的格局设计

13. 和式客房

14. 西式客房

13

14

Tsubaki

海石榴温泉旅馆

Location: 776 Miyakami, Yugawara-machi, Ashigarashimo-gun, Kanagawa, 259-0314, Japan

Spa Name: Yugawara Onsen

Area: 18,748 m²

Quantity of Guest Room: 29

Quantity of Spa Bath: 4

Other Facilities: guest room with open-air spa

地址：日本神奈川县足柄下郡汤河原町奥汤河原宫上776

温泉名：汤河原温泉

面积：18748平方米

客房数：29

温泉浴池数：4

其他设施：带露天浴池的客房

Nestled in Miyakami, Yugawara-machi, Ashigarashimo-gun, Kanagawa, this striking beauty is found within an hour from the heart of Tokyo by car. The Kabukimon, as well as the cottages of Tsubaki, are built in the Sukiya style, traditional architectural style – characterised by delicate sensibility and ornamented simplicity, in accordance with classic Japanese designing style of Sagano. All buildings are enclosed in Japanese gardens. Covering an area of more than 3000 square metres, the Geihin-kan Pavilion with only five luxurious guestrooms received the best prize of the Kanagawa-Prefecture Architectural Contest in 1985. The main gate, known as Kabukimon, welcomes the guest as a prelude to serene and peaceful visit. The hot spring comes from Yugawara Onsen, which has long been crowned as "Yakushi No Yu". The abundant flow of Tubaki Onsen forms permanent stream all seasons. Lounging in a spacious Grand Hot Spring Spa soothes away daily fatigue and stress. This is the traditional Japanese lifestyle and onsen culture.

Tsubaki is a "Ryotei-Ryokan", a long-history Japanese-style auberge, which features the traditional Japanese Kaiseki cuisine of Kyoto. Guests can enjoy the Kaiseki cuisine made of thin sheets of bean curd, endowed by nature, and then, lying in bed with a soothing heart after enjoying the Japanese food culture.

海石榴温泉旅馆位于神奈川县足柄下郡汤河原町奥汤河原，距东京1小时的车程，采用的是恪守日本古典传统的京都嵯峨野式建筑的整体设计理念，简洁、极富感性。木工建筑和材料引用了数寄屋风格的建筑方法。周围由日式庭园所包围。旅馆内名为"迎宾馆"的高级贵宾室曾在1985年获得神奈川县建筑设计大赛最优奖。迎宾馆占地3000多平方米，只有5间豪华客房。名为"歌舞伎门"的大门将客人迎至一个宁静平和的世界。

海石榴温泉为"汤河原温泉"。"汤河原温泉"自古代以来就有能治百病的"药师之汤"的美誉。海石榴温泉四季常流而且水量充沛。在温泉里放松全身可将游客的疲劳及烦恼一扫而光，这就是日式的洗身之道和温泉文化。

海石榴源自"料亭旅馆"，它继承了古都京都日本传统食文化"怀石料理"，颇具特色。在这里游客可以享受大自然的恩惠，品尝豆腐皮制成的怀石料理。美食之后静心入梦，海石榴就是这样的"料亭旅馆"。

1. The night view of the façade

2. The façade of the Kyo ryokan consulting the traditional Japanese classic designing style of Sagano

3. The indoor spa bath with permanent flow of hot spring all seasons

4. The VIP room – the Geihin-kan room with Koya Maki open-air spa

1. 外观夜景

2. 参照京都嵯峨野设计的京风旅馆外观

3. 室内温泉大浴池，温泉四季常流

4. 迎宾馆贵宾室，带专用高野槇露天温泉浴室

5

5. The VIP room of Geihin-kan

5. 迎宾馆贵宾室

3

4

6. TheJapanese-style guestroom with exclusive spa

7-8. The Geihin-kan room with exclusive open-air spa bath

6. 纯日式客房，专用露天温泉浴室

7－8. 迎宾馆贵宾房，专用露天温泉浴室

9. Zen Style SPA ichirin in which you can enjoy the utmost comfort

10. The salon bar "Shu-chu-ka" surrounded by gardens

11. The exclusive spa bath in the Japanese and Western-style guestroom

9. 禅意风格的 SPA 提供最舒适的享受

10. 四周庭园环绕的沙龙吧 "酒中花"

11. 客房，日式融合西式风格，配专用露天温泉浴室

12

12. The guest room in Japanese and Western style with exclusive open-air spa

13. The guest room of Geihin-kan

14. The Japanese and Western-style guest room

12. 日式融合西式风格的客房带有专用露天温泉浴室

13. 迎宾馆贵宾室

14. 日式融合西式风格的客房

Sansuirou

山翠楼温泉旅馆

Location: 673, Miyakami, Yugawara, Asigarasimogun, Kanagawa Prefecture, 259-0314, Japan

Spa Name: Yugawara Onsen

Area: 3,622 m²

Quantity of Guest Room: 57

Quantity of Spa Bath: 4

Other Facilities: guest room with open-air spa

地址：日本神奈川县足柄下郡汤河原町宫上673

温泉名：汤河原温泉

面积：3 622平方米

客房数：57

温泉浴池数：4

其他设施：带露天浴池的客房

Located in Yugawara, Asigarasimogun, Kanagawa Prefecture, Sansuirou and Tubaki are two hotels of the same company yet with their own characteristics. Sansuirou has been crowned as Mini Kyoto of Sagami since its foundation in Shouwa 8 (1933) and was chosen as one of the "250 Most Popular Onsen Ryokans with 5-star Accomation Facilities" in 2010. All of the four complexes "Jurakudai", "Jyobandai", the annexe "Genji" and "Momayamadai" of Sansuirou have a broad view for its high-level terrain (there are even open-air spa pool on the seventh floor), commanding a panoramic view of the green water of Fujiki River, while the starry sky is more remarkable as the stars are so close that is even reachable with hands,

Upon entering from the vestibule, then the reception desk and the corridor, guests are blessed with a surene and peaceful mind within the spacious room with gentle light. Each and every guest room is designed with their own view of landscapes such as varied and graceful trees and jungles, the beautiful scenery of Somma and the clear stream of Fujiki River. The hot spring here all come from Yugawara Onsen. There are not only indoor open-air spas but also the observatory large communal spa named "Ozora", making guests fully intoxicated in the amorous feelings endowed by nature. Moreover, wormwood

fumigation service is also available here. Guests can enjoy traditional Kaiseki cuisine in guest rooms of pure Japanese style.

山翠楼温泉旅馆是海石榴温泉旅馆的姊妹馆。该馆位于神奈川县足柄下郡汤河原町，始建于日本昭和八年（1933年），被誉为相模小京都。2010年入选"人气温泉旅馆酒店250选5星级住宿设施"，由4座独立茶室——"聚乐第"、"常盘第"、"源氏别馆"、"桃山第"等建筑群组成，地势高（甚至在8楼设有露天浴室）、视野开阔，箱根的群山、藤木川的绿水一览无余，夜晚的星空与人近得似乎伸手即可摘得。

山翠楼从玄关开始到服务前台、走廊通道，灯光柔和、空间宽敞，让人一走进"山翠楼"就产生恬静平和的心态 。山翠楼根据客房的不同方位而设计，让客人可以欣赏到多姿多彩的树木丛林：有的能眺望到箱根外轮山，有的能俯视藤木川清流。山翠楼温泉泉质为"汤河原温泉"。不仅客房附带露天温泉，还有露天大浴池"大空"让游人尽情体验大自然恩赐的万般风情。旅馆还提供艾草熏蒸服务。客人们可以在纯和式客房里享用日本传统"怀石料理"。

1. The façade of the hotel

2. The exclusive open-air spa of the luxury guest room "Momoyamadai"

3. The vestibule – the entrance

4. The corridor

1. 外观

2. 豪华客房"桃山第"的专用露天浴池

3. 玄关入口

4. 馆内走廊

5. The hall of the hotel in the style of Kyoto Ryotei Ryokan

5. 京都料亭风格的旅馆大厅

6. The observatory open-air spa "Ozora", with a commanding view of the natural scene of Okuyugawara

7. The exclusive open-air spa of the luxury guest room "Momoyamadai"

8. The indoor large bath in which you can admire the beautiful Japanese garden

6. 展望露天浴池"大空"，可眺望奥汤河原的大自然景色

7. 豪华客房"桃山第"的专用露天浴池

8. 室内温泉大浴池，入浴的同时可观赏到美丽的日式庭园

9. The beauty salon "Madoromi SPA SUI"

10. The exclusive terrace of the guest room "Jyobandai"

11. The Japanese-style guest room "Jyobandai"

9. 美容室

10. "聚乐第"客房的专用露台

11. 纯和式客房"聚乐第"，带有专用露天浴池

14

12. The dressing room of luxury guest room "Momoyamadai"

13. The luxury guest room "Momoyamadai", Japanese style with an area of 16 tatami mats (26 square metres)

14. The guest room named "Momoyamadai" with 16 tatami mats, Western-style bed room, a dressing room, an indoor bath pool and an open-air spa

12. 豪华客房"桃山第"化妆间

13. 豪华客房"桃山第"主间（纯和式），面积16帖榻榻米（26平方米）

14. 豪华客房"桃山第"，面积榻榻米16帖＋西式卧室＋化妆室＋室内浴池＋露天浴池

Seiransou

青峦庄温泉旅馆

Location: 679, Miyakami, Yugawara, Asigarasimogun, Kanagawa Prefecture, 259-0314, Japan

Spa Name: Yugawara Onsen

Quantity of Guest Room: 48

Quantity of Spa Bath: 5

Other Facilities: guest room with open-air spa, rentable spa

地址：日本神奈川县足柄下郡汤河原町宫上679

温泉名：汤河原温泉

客房数：48

温泉浴池数：5

其他设施：带温泉浴室的客房、包租温泉

Located in Yugawara, Asigarasimogun, which is only 90 minutes' walk from Tokyo, Seiransou is an historic and famous onsen hotel with a history of more than 80 years. With graceful natural scenery, Oku-Yugawara is a county park in Kanagawa Prefecture, which has a long history since its foundation in Shouwa Age, while many historical sites of that time can still be found now. The vermeil wooden bridge stretching across the opposite Hakuryu Waterfall, which falls from 38 metres above. Seiransou is characterised by a vivid picture of wild phenomena such as the wriggling Hakuryu Waterfall next to the guestrooms, the open-air spa surrounded by wonderful mountains and stones in odd shapes. Guests can sooth their hearts as a part of the green mountains and the blue water while listening to the artistic rippling song of the running water. It was the blue sky and the green mountains that gave birth to the name of the hotel.

The two hot springs here all comes from its own property. The water temperature is kept at 70 degree to 80 degree with a gushing volume of 100 litres per minute. The indoor spas also use the circulating hot spring. The water in the spa has chloride and vitoriol elements, which boasts the quality of alkalescence and hypotonicity. Transparent and free from extraneous odour, the hot water makes guests feel comfortable with its constant temperature, producing a significant effect in curing neuralgia, osphyalgia and gynaecopathia.

青峦庄温泉旅馆位于神奈川县足柄下郡汤河原町，距离东京90分钟路程，是一家有80余年历史的老字号温泉旅馆。奥汤河原是神奈川县的县立公园，自然风景优美，具有悠久的文化历史，至今仍保留着日本昭和期间建造的众多历史古迹。横跨河涧的朱红色木桥正对着落差38米的白龙之瀑。野趣横生是青峦庄温泉旅馆的最大特点，因为它给游人展示的是这样一个画卷：客房旁边的白龙瀑布蜿蜒而下，野天浴池被奇山异石包围，一边听着潺潺流水之声，一边将身心融入绿水青山之中。青蓝的天空加上绿色融融的山峦正是旅馆名字的由来。

青峦庄温泉旅馆所有2个温泉源泉，水温70~80度，每分钟涌出量达100升之多。温泉水质为盐化物硫酸盐泉，具弱碱性和低张性，无异味而透明，入水舒适，保持恒温。对神经疼痛、腰部疼痛、妇科病有特别功效。

1. The open-air spa commanding a view of the waterfall

2. The façade of the garden of the annexe

3. The paradise-like open-air spa

1. 露天温泉可同时眺望瀑布

2. 别馆庭院外观

3. 仙境般的露天温泉

4. The vermeil bridge to the open-air spa

4. 通往露天温泉的红桥

5. The open-air spa for women

6. The indoor spa for men

7. The paradise-like open-air spa

5. 女性专用露天浴池

6. 男士室内大浴池

7. 仙境般的露天温泉

8. The outdoor bathing place in the annexe (available 24 hours)

9. Women's spa (available 24 hours)

10. The banquet hall

8. 别馆中的浴场（24小时可使用）

9. 女性大浴场（24小时可使用）

10. 宴会厅

11. A special guest room in the annexe

12. The guest room in the annexe (16 square metres), Japanese-style guest room with its own bathroom

13. The guest room in the main building (16 square-metre), Japanese-style guest room with its own bathroom

11. 别馆特别客房

12. 别馆带独立卫浴的和式客房，16平方米

13. 本馆带独立卫生间的和式客房，16平方米

Onyadomegumi

御宿惠温泉旅馆

Location: 361 Miyakami, Yugawara-machi, Ashigarashimo-gun, Kanagawa Prefecture, 259-0314, Japan

Spa Name: Yugawara Onsen

Area: 2,848 m²

Quantity of Guest Room: 37

Quantity of Spa Bath: 4

Other Facilities: private open-air spa, 2 rentable spa, foot spa

地址：日本神奈川县足柄下郡汤河原町宫上361

温泉名：汤河原温泉

面积：2848平方米

客房数：37

温泉浴池数：4

其他设施：专属露天浴池、包租浴室2个、足汤

Enjoying the ideal location of the central area of Yugawara Onsen in Yugawara-machi, Ashigarashimo-gun, Kanagawa Prefecture and the convenient transportation, Onyadomegumi is a famous spa and health resort in Japan. To realise the concept of "Megumi" (affluence and happiness), it is a typical modern architecture combining the ancient and modern designing techniques. Besides its indurative shatter-proof structure, the main building "Seseragi-Kan" founded in 1995 and the annexe "Tsuki no takumi" also incorporates the merits of buildings in Edo era. A foot spa is designed in the atrium for guests to have a warm and comfortable foot spa. The women's open-air spa "Ebisu-No-Yu" is made of granite, whose tone and thickness demonstrate fully the artistic effect of the natural stone. The men's open-air spa "Bishamon-No-Yu", which was decorated by Libanotis buchtormensis made of granite and the bamboo curtain, looks extraordinarily simple and antique. Facing the garden and the mountains, the private chartered bath "Ginga" creates a feeling of openness, while evokes mysterious dream with its grand vision at night when plants and flowers in the garden are illuminated by lights.

Taking guests' diversified requirements into account, all the facilities here are barrier free design, which guarantee a comfort and cosy living environment for guests.

御宿惠温泉旅馆位于神奈川县足柄下郡汤河原町，地处汤河原温泉中心地带，交通十分便利，是日本著名的温泉疗养胜地。旅馆在设计上充分体现"惠"这一核心，1995年建成的本馆"涓涓馆"和2008年建成的别邸"月之匠"，采用坚固的耐震构造，同时融入日本江户时代建筑诸多特点，是一座典型的古今结合的现代建筑。御宿惠温泉旅馆在中庭特别设计了足汤，让游人疲惫的双脚感受温和的足汤浴。女士露天温泉"惠比寿之汤"采用御影石材，色调及厚度感充分体现出天然石质的美观效果；男士露天温泉"毗沙门之汤"采用御影石材配竹围帐，显得格外古朴。包租温泉浴室"吟雅"面朝庭园和山间，给人以开放感。夜晚庭园内灯光映衬花木，尽显繁华景象，让人产生神秘遐想。

御宿惠温泉旅馆采用无障碍设计，充分考虑客人的多样化需求，确保游客住得"惠"心、舒适。

1. The foot spa designed uniquely in the atrium, feet relax in the gentle hot spring while admiring the flowers and plants in the midnight light, the wonderful garden scenery

2. The façade of the tile-roofed building, designed with the prototype of shops in Edo period

3. Men's open-air spa "Bishamon-No-Yu", made of andesite

4. Women's open-air spa "Ebisu-No-Yu", made of andesite, whose tone and thickness demonstrate fully the artistic effect of natural stone

1. 中庭特别设计的足汤让疲惫的双脚感受温和，夜晚庭园灯光映衬花木也非常值得一看

2. 外观，日本瓦房顶建筑，仿江户时代的商店设计

3. 男士露天温泉"毗沙门之汤"采用御影石材砌成

4. 女士露天温泉"惠比寿之汤"采用御影石材，色调及厚度感充分体现出天然石质的美观效果

5. The spacious hall with historic taste

5. 大厅宽敞而古朴

1. 中庭特别设计的足汤让疲惫的双脚感受温和，夜晚庭园灯光映衬花木也非常值得一看

2. 外观，日本瓦房顶建筑，仿江户时代的商店设计

3. 男士露天温泉"毗沙门之汤"采用御影石材砌成

4. 女士露天温泉"惠比寿之汤"采用御影石材，色调及厚度感充分体现出天然石质的美观效果

3

4

6

6. Men's indoor spa "Bishamon-No-Yu" with high-grade cypress as the decorative material for fringe

7. Women's indoor spa "Ebisu-No-Yu" is made of andesite

8. The exclusive open-air spa in the Japanese-style guest room

6. 男士室内大浴池"毗沙门之汤"，浴池边缘采用高级桧木

7. 女士室内大浴池"惠比寿之汤"，浴池采用铁平石

8. 带有专用露天温泉的日式客房

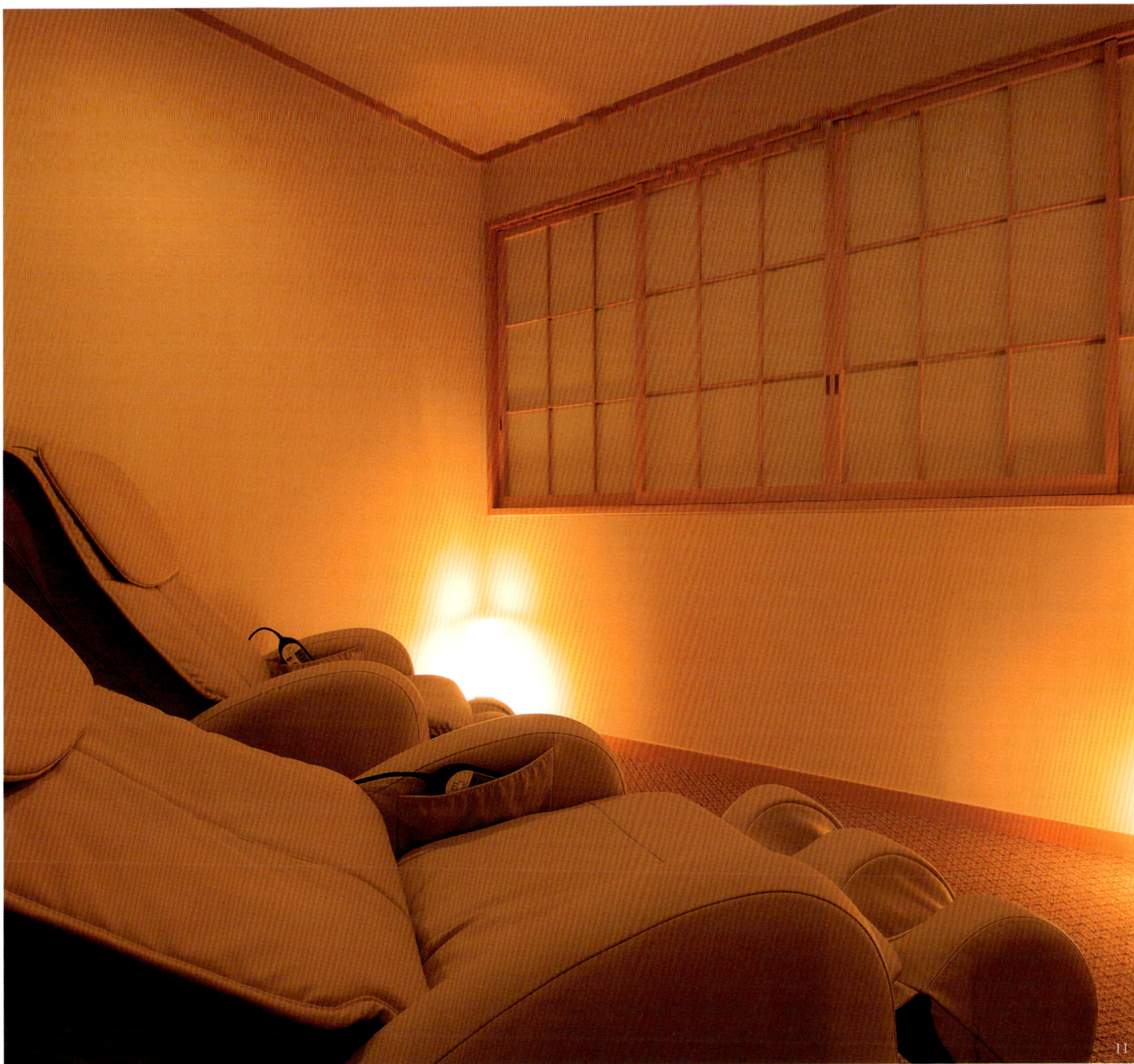

9. The chartered spa bath "Ginga" facing the garden and the mountain with a sense of openness

10. The panorama of the Japanese and Western-style guestroom with exclusive open-air spa on the outdoor terrace, covering a total area of 87 square metres

11. The rest area equipped with two massage armchairs

9. 包租温泉浴室"吟雅",面朝庭院和山间,有开放感

10. 日式融合西式风格的客房全景,客房总面积为87平方米

11. 休憩间,配有2台按摩椅可免费使用

12. The Japanese-style guest room with exclusive open-air spa, made of tuff

13. The dinning room in the Japanese-style guest room with exclusive open-air spa

14. The Japanese and Western-style guest room with exclusive open-air spa on the outdoor terrace, covering a total area of 67 square metres

12. 带有专用露天温泉的日式客房，浴池采用铁平石

13. 带有专用露天温泉的日式客房的餐厅间

14. 日式融合西式风格的客房，露天温泉设在室外露台，客房总面积有67平方米

13

14

Mikawaya Ryokan

三河屋温泉旅馆

Location: 503 Kowakidani, Hakone-machi, Ashigarashimo-gun, Kanagawa, 250-0406, Japan

Spa Name: Kowakidani Onsen

Area: 66,000 m²

Quantity of Guest Room: 35

Quantity of Spa Bath: 6

Other Facilities: guest room with indoor open-air spa, rentable spa

地址：日本神奈川县足柄下郡箱根小涌谷503

温泉名：箱根小涌谷温泉

面积：66000平方米

客房数：35

温泉浴池数：6

其他设施：带露天温泉浴室的客房、包租浴室

Seated in Kowakidani Mountain, Hakone-machi, Ashigarashimo-gun, Kanagawa, approximately 660 metres above sea level, Mikawaya Ryokan is an old brand with a building history of more than 10 decades since its foundation in Meiji 16 (1883 AD). It is not only an elegant and graceful hotel, the central meaning of its theme "Fuga", but also a resort full of rich historical stories. The great pioneer of the democratic revolution of China, Sun Yat-sen has enjoyed his stay here in the guest room named "Shochiku", where the writing of Mr. Sun's "High Mountains and Floating Water" is still carefully kept here. It is also favoured by many Japanese poets and writers, among whom the famous Japanese drawer Yumeji Takehisa left his great work "Tanabata" (The Star Festival), which is hung at the vestibule, after enjoying his stay here. Moreover, it is the refuge of numerous superstars, filmstars and writers. The lounge in the style of Taisho period can still be seen in the hotel.

The guest rooms with open-air spa named "Kasumi" and "Tsutsuji" put together all graceful elements of the traditional Japanese ryokan. The large bath "Meji", with a long history and culture, inherits the romantic charm of Meji period. Guests can enjoy the bathing while admiring the grand scenery of the ever-changing natural landscape in different seasons: sakura and cuckoo in spring, lilium brownii in summer, red maple leaves in autumn and the snow-covered landscape in winter – the Eastern way of life that men and the nature live in harmony is incisively and vividly demonstrated here. The simple alkaline hot spring has a significant curative effect in skincare with little mineral substance and less durability.

三河屋温泉旅馆地处神奈川县足柄下郡箱根小涌谷山中，海拔660米，创建于明治十六年（1883年），是一座有着一百多年历史的老店。旅馆的设计主题为"风雅"，不仅"风雅"，还散发着浓郁的历史气息。中国民主革命的伟大先驱孙中山先生就曾在旅馆的"松竹"屋下榻，客房保留了孙先生的墨宝"山水清幽"。三河屋温泉旅馆深受日本文人墨客喜爱，日本著名画家竹久梦二不仅在此下榻，而且还留有作品《七夕》挂在玄关处。歌星、影星、作家更是络绎不绝。游客在这里还能见到属于大正时代的建筑物——如今是旅馆的休息室。

三河屋温泉旅馆设有带露天浴池的客房，如"霞馆"和"杜鹃"等，集中了日本传统旅馆的各种优美元素。"明治"大浴室、露天浴池传承了明治时代的神韵，历史弥久、源泉弥新。游客一边沐浴一边欣赏明神岳山、大文字山、河浅间山的雄伟美景。这里景色四时变幻，春有樱花和杜鹃，夏有山百合，秋看枫叶，冬观白雪，人与自然相互依存的东方生活观在这里体现得淋漓尽致。温泉泉质为碱性单纯温泉，矿物质等含量较低，持久性小，美肤功效甚佳。

1. The outdoor view which can be overlooked from the guest room (in April), the broad garden in the simple and elegant colour of cherry blossom

2. The garden in summer – as the affiliated garden, Houraien boasts the enchanting scenery changing with seasons.

3. The façade of the Japanese-style hotel with a history of 10 decades

1. 从客房可眺望到的景色，每到4月，宽敞的庭园被淡雅的樱色渲染

2. 庭园，夏。蓬莱园是其附属庭园，园内景致迷人，四季变幻

3. 百年历史的纯和风旅馆外观

4. The façade

4. 外观

5. The vestibule permeated with dense history breath

6. The corridor inside

5. 玄关洋溢着浓浓的历史气息

6. 馆内走廊

7. The independent guest room "Hanare", the guest room "TSUTSUJI" with open-air spa on the terrace and a private garden with blooming sakura (the first and middle ten days of April)

8. The large bath "Meiji"

7. 带有露天温泉的独立客房"杜鹃"。露台上设有露天浴池，专属庭园中樱花盛开（4月上旬至中旬）

8. 大浴室"明治"

9. The large bath "Meiji"

10. Men's open-air spa

11. The stairs to the guestroom "Sho chiku"

9. 大浴室 "明治"

10. 男士露天浴池

11. 通往客室 "松竹" 的楼梯

14

12. The guest room "Sho chiku", where you can overlook the outdoor garden on the terrace

13. The guest room "Sho chiku", where the great Chinese leader Sun Yat-sen had stayed

14. The guest room with open-air spa named "Kasumi"

12. 客房"松竹"，从露台可以一望室外庭园

13. 客房"松竹"是中国伟人孙中山曾留宿的客房

14. "霞馆"，带露天温泉的客房

Hakone Hotel Kowakien

箱根小涌园温泉饭店

Location: 1297 Ninotaira Hakone-machi, Ashigarashimo-gun, Kanagawa-ken, 250-0407, Japna

Spa Name: Kowakidani Onsen

Area: 16,000 m²

Quantity of Guest Room: 220

Quantity of Spa Bath: 4

地址：日本神奈川县足柄下郡箱根町二平1297

温泉名：箱根小涌谷温泉

面积：1.6万平方米

客房数：220

温泉浴池数：4

Located in Ninotaira Hakone-machi, Ashigarashimo-gun, Kanagawa-ken, the most popular tourist attraction in Japan, Hakone Hotel Kowakien enjoys a long history and moreover wins the great affection from people at home and abroad in that its hot spring (spring water is rich in beneficial mineral substances), as gentle as astringent, has the special efficacy to skincare. Covering an area of 16,000 square metres, it is a recreation centre focusing on the theme of onsen. There are various kinds of baths within the five onsen areas with their own themes such as the wine onsen, the coffee onsen, the most outstanding one – the miraculous Dead Sea Onsen and the traditional Japanese onsen "Mori no Yu", which combines the magnificence of Hakone and the glamour of nature together. Surrounded by beautiful natural scenery, "Mori No Yu" is the largest open-air spa in Hakone, whose marvelous space design affords guests a panoramic view of the vasty garden and the splendid scenery of Hakone Somma.

Appetising dishes from across the world are served in the five dinning rooms of the hotel, such as Japanese Kaiseki Cuisine, the superior Mediterranean Cuisine, Teppangaki and the like.

箱根小涌园温泉饭店位于神奈川县足柄下郡箱根町，地处箱根——日本最著名的旅游胜地，它不仅历史悠久，而且因其泉质如化妆水般温柔并有特殊的美肤功效（水质富含有益矿物质），所以深受各国游客的喜爱。箱根小涌园温泉饭店是以温泉为主题的休闲娱乐中心，占地1.6万平方米 。饭店园区内有五大温泉主题设施，提供多种泡汤方式，泳装区的葡萄酒温泉、咖啡温泉、神奇的"死海温泉"是这里的特色。此外，还有融合了箱根的雄伟和自然美景于一体的日式传统温泉"森之汤"。"森之汤"也是箱根最大的露天温泉浴池，四周美景如画。饭店内的空间设计美轮美奂，并且从每一间客房都可以看到室外庭园和箱根外轮山的壮丽风光。

饭店内有5个餐厅，提供美味可口的各国佳肴，如日式怀石料理、绝佳的地中海式料理、铁板烧等。

1. The night scene of the façade

2. The open-air spa "Mori no Yu", the largest hot spring bath in Hakone

3. The open-air wine spa

4. The standard twin room – there are so many accommodation choices here with considerate services from the standard double-bed room to the elegant royal suites

1. 外观夜景

2. 露天温泉 "森之汤" 是箱根最大的露天温泉浴池

3. 露天葡萄酒温泉

4.标准双人间。这里提供从标准双人间到典雅的皇家套房等多种住宿选择，服务周到

5. The terrace of the guest room, which commands a panoramic view of the broad garden and the splendid landscape of Hakone Somma

5. 客房露台。每间客房都可以看到庭园美景和箱根外轮山的壮丽风光

Hotel ChooBoo Sansou

眺望山庄温泉酒店

Location: 698-25 Miyashita, Yugawara, Asigarasimogun, Kanagawa Prefecture, 259-0304, Japan

Spa Name: Yugawara Onsen

Area: 9,000 m²

Quantity of Guest Room: 6

Quantity of Spa Bath: 3

Other Facilities: rentable spa

地址：日本神奈川县足柄下郡汤河原町宫下698－25

温泉名：汤河原温泉

面积：9000平方米

客房数：6

温泉浴池数：3

其他设施：包租浴室

Built on the hathpace near Yugawara Station, Hotel ChooBoo enjoys the ideal location in Yugawara Onsen Town in Yugawara, Asigarasimogun, Kanagawa Prefecture. Taking full advantage of the ideal geographical position with a long history and particularly favourable natural condition, it creates a surreal world for guests, being a part of nature. Just as the meaning implied by the hotel name, the designer aimed to build a hotel on the hathpace, which has a commanding view of infinite scenes. Guests can enjoy the cosy and comfortable onsens here while more importantly, overlook the sky, blue sea, the lush forest on mountains far away, the pretty neighbourhood and the winding Hydra-like Shinkansen, feeling like being integrated in the sky. The typical Western and Japanese-style buildings against the background of all these wonderful scenes make guests feel better than the lively supernatural beings.

The natural alkaline hot spring has the bubble massage function. There are various kinds of cuisine served here, the Western style and the Japanese style, which can be chosen by personal habit.

眺望山庄温泉酒店位于神奈川县足柄下郡汤河原町，地处日本著名的温泉胜地汤河原温泉区，借着汤河原车站附近的高台依山而建，历史久远、风光秀丽。眺望山庄温泉酒店充分利用这一得天独厚的地理位置，为游人创造一个融入大自然的超脱境界。眺望山庄温泉酒店充分表达了设计者的思想，温泉浴池建于高台，将无限风光尽收眼底。在这里，游人不仅可以享受到温泉的舒适与惬意，更可以极目远眺，蔚蓝色的大海、远山的密林、美丽的街区、蜿蜒长蛇般的新干线尽收眼底，让人有一种融入蓝天的感觉。酒店建筑设计融合了典型西洋风格与和式风格，更让游人体会到赛过活神仙的生活享受。

酒店的温泉泉质为天然弱碱性，并带有水泡按摩功能。眺望山庄温泉酒店为客人准备的日式料理和西餐，可以依个人喜好选择。

1. The façade

2. The open-air spa

3. The surrounding scenic spots

4. The large spa pool with the function of bubble massage

1. 外观

2. 露天温泉浴池

3. 周边景点

4. 温泉大浴池，带水泡按摩功能

5

5 The surrounding scenic spots

5. 周边景点

3

4

6. The large spa with the function of bubble massage

7. The guest room in unique style

8. The double-bed room in Western style, with a panoramic view of the seascape

6. 露天温泉浴池位于高处，视野开阔，风景独好

7. 风格独特的客房

8. 双人西式客房，窗外海景一览无余

Yugawara Mizunokaori

汤河原水之香里温泉酒店

Location: 614 Miyakami, Yugawara-machi, Ashigarashimo-gun, Kanagawa Prefecture, 259-0314, Japan

Spa Name: Yugawara Onsen

Area: 1,421 m²

Quantity of Guest Room: 25

Quantity of Spa Bath: 4

Other Facilities: private open-air spa

地址：日本神奈川县足柄下郡汤河原町宫上614

温泉名：汤河原温泉

面积：1421平方米

客房数：25

温泉浴池数：4

其他设施：私人专属露天浴池

Nestled in Miyakami, Yugawara-machi, Ashigarashimo-gun, Kanagawa Prefecture, it was built along the river with rippling water and rolling stream. The hot spring, coming from its own properties, can make skin soft, moist and smooth, and the gentle water quality without any irritating substances is suitable for children. Guests can soothe their hearts in the bright and spacious large spa in their favourite bathrobes. However, the open-air spa endow guests with more interests and pleasure. From the early morning to the midnight, guests can feel the same natural beauty and amorous feelings but different joyness brought by the changing time. The standard guest room with an area of 17 square metres has a commanding view beautiful scenes of Yugawara – the picturesque overlapping mountains and green trees, tranquil and peaceful.

It is only takes 10 minutes to walk from the hotel to Manyo Park, where guests enjoying foot spa while admiring diversified sceneries in four seasons such as the fireflies all over the garden in summer and the fiery-red maple leaves in autumn.

The Japanese Kaiseki Cuisine, made of fresh seasonal ingredients, has won good reputation among guests. There is also delicate coffee for guests. All the thoughtful service leaves guests unforgettable memories.

水之香里温泉酒店位于神奈川县足柄下郡汤河原町宫上，依河而建，伴随潺潺流水奔流不息。水之香里温泉使用的泉水为自家源泉涌出，室内设有岩石浴室。泉水落在肌肤上柔润顺滑，没有任何刺激性物质，很适宜小孩入浴。大浴场宽敞明亮，客人可以选择穿着自己喜欢的浴衣沐浴，使得沐浴更有情趣。无论清晨白昼、夜晚三更，客人们在露天温泉浴池都能感受周围的自然风情，享受时间变换的乐趣。标准日式客房约17平方米，从客房极目远眺汤河原的美景，远山叠翠如诗如画，安详恬静。

从酒店步行10分钟就到了万叶公园，在公园里可以进行足浴，夏季还可以欣赏到满园萤火虫飞舞，而秋季又可观赏火红的枫叶，四季皆有美景变化万千。

水之香里温泉酒店采用新鲜时令食材制作的和式会席膳深受客人好评。酒店大厅还为客人准备了美味的免费咖啡，周到的服务给游客留下难忘的记忆。

1. The façade of the hotel with a creek trickling sluggishly nearby

2. Gentlemen's observatory open-air spa where you can overlook the stunning scenery of Yugawara while enjoying birds' tweedle

3. Ladies' observatory open-air spa within the environment of rolling and overlapping mountains with lush trees

4. The restaurant. Guest can choose to dine in the restaurant or in the guest room

1. 酒店外观，附近一条小河涓涓流过

2. 男士露天浴池，可远眺汤河原美景，聆听小鸟歌唱

3. 女士露天浴池，远山叠翠，延绵不绝

4. 餐厅，客人也可选择在客房用餐

5. The indoor spa, the rock bathroom

5. 室内大浴池、岩浴室

6. The standard Japanese-style guest room with an area of 17 square metres, where you can enjoy the natural beauty and tranquility of Yugawara

7. The barrier free guest room designed for the elder and disabled

6. 标准日式客房，面积约17平方米，可享受汤河原大自然的美妙与恬静

7. 无障碍客房，专为伤残或高龄住客设计

Hotel Shiroyama

城山温泉酒店

Location: 207 Shirohori, Yugawara-machi, Ashigarashimo-gun, Kanagawa Prefecture, 259-0314, Japan

Spa Name: Yugawara Onsen

Area: 1,152.42 m²

Quantity of Guest Room: 20

Quantity of Spa Bath: 4

Other Facilities: guest room with open-air spa, rentable spa

地址：日本神奈川县足柄下郡汤河原城堀207

温泉名：汤河原温泉

面积：1152.42平方米

客房数：20

温泉浴池数：4

其他设施：带露天温泉的客房、包租温泉

It is only two minutes' walk from Yugawara Tram Stop to Hotel Shiroyama, located in Shirohori, Yugawara-machi, Ashigarashimo-gun, Kanagawa Prefecture. Hotel Shiroyama and Yugawara Mizunokaori are two chain hotels with different names and characteristics. Yugawara Onsen can be divided into two types according to the different qualities: the meek and smooth natural hot spring and the hot spring with rich minerals. To meet guests' requirements of multiaspects, some guest rooms underwent redecoration in July, 2007 when the Japanese-style guest rooms with open-air spa, Japanese and Western-style guest rooms, double-bed rooms were added. There are also large Japanese-style rooms which can accommodate eight persons, guest rooms with open-air spa named "Aoi" and "Koutou" and the Western-style guest rooms named "Shinonome".

The restaurant features seasonal "Hana Kaiseki Zen", cooked with the fresh marine fish from the nearby sea and local vegetables and fruits. Moreover, guests can choose their own favoured menus such as the roast meat cuisine and the Kaiseki Ryori with abalone and Ise lobster. The family guests or travelling couples can enjoy their dinner with leisure in their rooms.

城山温泉酒店位于神奈川县足柄下郡汤河原城堀，距离汤河原电车站只需步行2分钟。城山温泉酒店与水之香里温泉酒店为姊妹馆。汤河原温泉的泉质主要分两种：柔顺滑润的天然温泉和富含矿物质的泉源。为了满足客人多方位的需求，2010年7月，酒店的一部分客房经过重新装修，增设了带露天温泉浴池的和式客房、日式西式融合客房、双人客房、能同时容纳8位客人的大面积和式客房、带露天温泉的客房"蓝"、"香橙"、东云西式客房。

酒店为客人提供时令感十足的"花会席膳"，使用的食材主要取自近海的新鲜海鱼和当地的蔬菜水果。客人可以根据自己的喜好，选择烤肉席膳、有鲍鱼和伊势龙虾的和会席膳等。因为晚餐是在客房内供应的，所以家族或者旅伴可以悠然自得地进餐。

1. The rentable open-air spa

2. The open-air spa of the guest room "Koutou"

3-4. The spacious Japanese-style guest room which can accommodate eight persons

1. 包租露天温泉

2. 客房"香橙"的露天温泉浴室

3－4. 宽敞的和式客房，可同时容纳8人住宿

5

5. The guest room "Aoi" with open-air spa

6. The guest room "Koutou" with open-air spa

7. The Western-style guest room named "Shinonome"

5. 带露天温泉的客房"蓝"

6. 带露天温泉的客房"香橙"

7. 西式客房"东云"

Prince Hotel Manza

万座王子温泉大饭店

Location: Manza Onsen, Tsumagoi-mura, Agatsumagun, Gunma Prefecture, 377-1595, Japan

Spa Name: Manza Onsen

Area: 34,957 m²

Quantity of Guest Room: 232

Quantity of Spa Bath: open-air spa 6, indoor spa 4

地址：日本群马县吾妻郡嬬恋村万座温泉

温泉名：万座温泉

面积：34957平方米

客房数：232

温泉浴池数：露天浴池6个、室内大浴池4个

Prince Hotel Manza is located in Tsumagoi-mura Resorts, Agatsumagun, Gunma Prefecture. Manza Onsen has been crowned as "Onsen Beyond the Clouds" and "Onsen Closest to the Starry Sky", as it is the highest onsen that can be reached by trolley bus, located on Mt. Shirane, 1,800 metres above sea level. Manza Onsen is one of the best onsens in Japan for its ideal location in the environment the same like the three terrains with the longest longevity in the world. It is also worldwide-known for its abundant hot spring volume. The 27 kinds of minerals and trace elements contained in the milky white water have many effects, which have been described as the catholicon that can cure all diseases and has been favoured by women for its efficacy in skincare. Available 24 hours, there are women's exclusive spas, men's exclusive spas and the mixed ones. In the spacious onsen named "Komasaku" with the thorough openness, guests can experience various feelings in different seasons such as the green and fresh spring, the cool summer with gentle breeze, the affluent autumn with red maple leaves and the silver white winter. Guests may also soothe their bodies and mind by enjoying the two kinds of natural spas in "Ishiniwa Open-air Spa" – the white turbid hot spring spa and the yellow turbid hot spring spa, on the highland in the starry sky.

万座王子温泉大饭店位于群马县吾妻郡嬬恋村度假村，万座温泉坐落在白根山海拔1800米处，是在日本坐车能到达的海拔最高的温泉，有"浮在云端的温泉"和"离星空最近的温泉"的美誉。万座温泉地处世界三大长寿地带中，是日本首屈一指的高山温泉乡之一。同时，它还以水量丰富闻名遐迩，其乳白色温泉水含有27种矿物质和微量元素，具有多种疗效，自古以来留下很多"医治百病"的佳话，更拥有令女性青睐的美肤效果。游客可以互用两家温泉设施，万座温泉可24小时享用。有女性专用、男性专用和男女共用的露天温泉。在宽敞的"驹草之汤"温泉里享受新绿之春、凉风之夏、红叶之秋和银白之冬，同样的温泉有不同的情趣。而在"石庭露天温泉"，游客更可享受两种不同的温泉泡汤乐趣——白浊温泉与黄混温泉，让游客在星星点点的高原夜空下，给心情和身体彻底放假。

こまくさの湯3
男女共用

1. The façade of the hotel in autumn

2. The autumn scenery of the open-air spa "Komasaku" of Prince Hotel Manza

3. The open-air spa "Komakusa" of Prince Hotel Manza

4. The summer scenery of "Ishiniwa Open-air Spa" of Prince Hotel Manza

5. The open-air spa "Komakusa" of Prince Hotel Manza – the top Takayama hot spring locates 1,800 metres high above sea level, which is called "Onsen Beyond the Clouds"

1. 外观，秋色

2. 秋天景色

3. 万座王子大饭店露天温泉"驹草之汤"

4. 石庭露天温泉，夏

5. "驹草（komasaku）之汤"，日本首屈一指的高山温泉，海拔1800米，被称为"浮在云端的温泉"

6. The winter scenery of the open-air spa "Komasaku"

6. 露天温泉冬天景色

標高一、八〇〇M 雲上の露天風呂

7. The night scene of "Ishiniwa Open-air Spa" of Prince Hotel Manza

8. The beauty salon, SPA

9. The Western-style guest room

7. 万座高原饭店的石庭露天温泉，夜景

8. 美容室，SPA

9. 西式风格的客房

Index 索引

Kanagawa Prefecture
Hakone Onsen

1. Mikawaya Ryokan
   503 Kowakidani, Hakone-machi, Ashigarashimo-gun, Kanagawa Prefecture, 250-0406, Japan
   http://www.hakone-mikawaya.com/

2. Gora Tensui
   1320-276 Gora, Hakone, Asigarasimo-gun, Kanagawa Prefecture, Japan
   http://www.gora-tensui.com/

3. Hotel Kajikaso
   688 Yumoto, Hakone-machi, Ashigarashimo-gun, Kanagawa Prefecture, Japan
   http://www.kajikaso.co.jp/

4. Yushintei
   193 Yumoto, Ashigarashimo-gun, Kanagawa Prefecture, Japan
   http://www.yushintei.co.jp/

5. Hakone Hougetu
   90-42 Hakone, Hakone, Asigarasimo-gun, Kanagawa Prefecture, 250-0522, Japan
   http://hakone-hougetu.com/

6. The Prince Hakone
   144 Motonhakone, Hakone-machi, Ashigara-shimo-gun, Kanagawa Prefecture, 250-0592, Japan
   http://www.princehotels.co.jp/the_prince_hakone

7. Hakone Hotel Kowakien
   1297 Ninotaira Hakone-machi, Ashigarashimo-gun, Kanagawa Prefecture, 250-0407, Japan
   http://www.hakoneho-kowakien.com/

Yugawara Onsen

8. Tubaki
   776 Miyakami, Yugawara-machi, Ashigarashimo-gun, Kanagawa Prefecture, 259-0314, Japan
   http://www.tubaki.net/

9. Sansuirou
   673, Miyakami, Yugawara, Asigarasimo-gun, Kanagawa Prefecture, 259-0314, Japan
   http://www.sansuirou.co.jp/

10. Hotel ChooBoo
    698-25 Miyashita, Yugawara, Asigarasimo-gun, Kanagawa Prefecture, 259-0304, Japan
    http://www.chooboo.co.jp/

11. Onyadomegumi
    361 Miyakami, Yugawara-machi, Ashigarashimo-gun, Kanagawa Prefecture, 259-0314, Japan
    http://www.onyadomegumi.co.jp/

12. Ootaki Hotel
    750-1 Miyakami, Yugawara-machi, Ashigarashimo-gun, Kanagawa Prefecture, 259-0314, Japan
    http://www.ootaki-hotel.com/

13. Yugawara Mizunokaori
    614 Miyakami, Yugawara-machi, Ashigarashimo-gun, Kanagawa Prefecture, 259-0314, Japan
    http://www.mizunokaori.com/

14. Hotel Shiroyama
    207 Shirohori, Yugawara-machi, Ashigarashimo-gun, Kanagawa Prefecture, 259-0314, Japan
    http://www.hotel-shiroyama.com/

15. Seiransou
    679, Miyakami, Yugawara, Asigarasimogun, Kanagawa Prefecture, 259-0314, Japan
    http://www.seiransou.co.jp/

Shizuoka Prefecture
Atami Onsen

16. Horai
    750-6 Mount Ito, Atami, Shizuoka Prefecture, 413-0002, Japan
    http://www.izusan-horai.com

17. Otsuki Hotel Wafuukan
    3-19 Higashikaigancho, Atami, Shizuoka Prefecture,413-0012, Japan
    http://www.wafuukan.com/

18. Hotel MICURAS
    3-19, Higashikaigancho, Atami, Shizuoka Prefecture, 413-0012, Japan
    http://www.micuras.jp/ http://www.orix.co.jp/grp/content/070119_MicurasJ.pdf

19. Atami Sakuraya Ryokan
    9-11 Higashikaigancho, Atami-shi, Shizuoka Prefecture, 413-0012, Japan
    http://www.atami-sakuraya.com/

20. Sun Hatoya
    572-12, Yukawa, Ito City, Shizuoka Prefecture, 414-0002, Japan
    http://www.sunhatoya.co.jp/home/index.html

21. Hotel Hatoya
    1391 Oka, Ito City, Shizuoka Prefecture 414-0055, Japan
    http://www.hatoyahotel.com/home/

22. Laforet Club Hotel Yitou
    2-3-1 Shishido, Ito City, Shizuoka Prefecture, 414-0004, Japan
    http://www.laforet.co.jp/lfhotels/ito/

23. Yokikan
    2-24, Suehiro-cho, Ito City, Shizuoka Prefecture, 414-0015, Japan
    http://www.yokikan.co.jp

24. IDUMISOU
    2-21 Okahiro-cho, Ito City, Shizuoka Prefecture, 414-0016, Japan
    http://www.idumisou.co.jp

25. ANGINE
    5-12 Nagisa-cho, Ito City, Shizuoka Prefecture, 414-0023, Japan
    http://www.angine.jp

26. Zagyosoh
    1741, Yawatano, Ito City, Shizuoka Prefecture, Japan
    http://www.zagyosoh.com/index.html

27. HANAFUBUKI Ryokan
    1041 Yawata, Ito City, Shizuoka Prefecture,413-0232, Japan
    http://www.hanafubuki.co.jp/kantai/index.htm

28. Southern Cross Resort Hotel
    1006 Yoshida, Ito City, Shizuoka Prefecture, 414-0051, Japan
    http://www.southerncross.co.jp/

29. Seizanyamato
    203 Oka, Ito-City Shizouka Prefecture, 414-0055, Japan
    http://www.seizanyamato.jp/